Man Environment
Reference 2

*Camera-ready copy for this volume
has been provided by the author.*

Man Environment Reference 2
Environmental Abstracts

Leon A. Pastalan
Editor

Daniel D. Jilek
Kent F. Spreckelmeyer
Stephen F. Verderber
Associate Editors

Debra S. Bogi
Rochelle Martin
Brian Lutz
Ruth E. Phillips
Ruth A. Tabler
Assistants

The University of Michigan Press
Ann Arbor

Acknowledgments

A special acknowledgment is made to Dean
Robert C. Metcalf of the College of Architecture
and Urban Planning and to Harold R. Johnson,
Co-Director of the Institute of Gerontology, for
providing financial and human resources that
made possible the completion of this reference
work. A debt of gratitude to Debra S. Bogi who
spent many weekends typing this manuscript.

Published in the United States of America by
The University of Michigan Press and simultaneously
in Rexdale, Canada, by John Wiley & Sons Canada, Limited
Manufactured in the United States of America

Library of Congress Cataloging in Publication Data
Main entry under title:

Man environment reference environmental abstracts.

　　Includes indexes.
　　1. Environment psychology — Abstracts. 2. Man —
Influence of environment — Abstracts. I. Pastalan,
Leon A., 1930 —
BF353.M34　　1983　　　155.9　　　83-5950
ISBN 0-472-08040-7

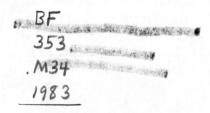

Contents

Preface

In gathering citations, we have attempted to select those studies and discussions that deal directly with the interface between the man-made environment and human behavior. Such documents represent the widest possible range of disciplines and fields of inquiry. Some come directly from the newly coalescing areas of environment psychology and man-environment studies, others from the behavioral sciences, from the planning and design fields, and the area of applied technology.

It is our feeling, given the very recent and multi-disciplinary nature of much of this literature, that it is essential, albeit extremely difficult, to organize this material within some sort of theoretical structure. It has been suggested by a number of individuals, and demonstrated by others, that these newly coalescing comprehensive areas are still in a pre-paradigmatic phase, and that any attempts to formulate a comprehensive theoretical framework would seem premature. So in at least partial response to this need, we are attempting to isolate what we characterize as 'relational concepts' -- which appear to be important variables in a range of interactions between man and his environment.

We also feel, given this diverse body of literature, that users of the book will desire to assess the information in specific ways reflecting particular interests or questions. Hence, the literature is organized along four dimensions -- the four dimensions which are felt to be the necessary and sufficient variables for characterizing a man-environment interaction. In this way a user of the book will be able to select citations in terms of specific populations being discussed or dealt with (along such dimensions as age, socio-economic status or cultural background), and in terms of particular relational concepts (privacy, complexity, scale).

It is hoped that such a series of selective indices will permit the book to be useful to the widest range of users -- the practitioner interested in a specific setting or population group; the researcher dealing with more conceptual issues; or any individual interested in exploring and thinking about the complex relationships between man's environment and that person's behavior.

The abstacts are selected on the basis of their representativeness in terms of focus and subject matter. The citations represent literally all the disciplines which either directly or indirectly are concerned with the relationships between man's environment and his or her behavior.

The mechanics of writing up the specific abstracts involve a standardized format. Each abstract contains the name of the author and the year of publication, the particular environmental setting in question along with the type or person of persons involved in the setting, elaboration of relational concepts regarding man and his environment and a brief statement as to the outcome(s) of the investigation or concern.

How to Use MER

MER consists of two major parts, Abstracts and Indexes.

ABSTRACTS: MER includes abstracts of 263 research based articles. The abstracts are arranged alphabetically by author's last name and each is assigned an Abstract Number (1-263). Each abstract also includes complete bibliographic information for the original article.

INDEXES: MER includes a dual indexing system providing the use with more specific information than does the traditional subject index.

Composite Index

Each abstract is categorized along four dimensions:

° Population studied
° Environment studied
° Relational concepts
° Behavioral outcomes

The composite index indicates all appropriate "key words" for each abstract.

AUTHOR YEAR	ABSTRACT NO.	MAN	SETTING	CONCEPTS	RELATIONAL OUTCOMES
AIELLO, J.R. & T.D. AIELLO, 1974	1	Children	Schools	Proxemics	Social Behavior
AIELLO, J.R., D.T. DeRISI, Y.M. EPSTEIN & R.A. KARLIN, 1977	2	Female students	Labora-tory	Crowding	Stress

KEY WORDS for the four dimensions are indexed separately. Key words are organized alphabetically, indicating appropriate abstract numbers.

Key Word Index

KEY WORD	ABSTRACT NUMBER
Children	1, 25, 29, 35, 53, 105, 107, 164, 165, 184, 186, 196, 214, 229, 259

"The Development of Personal Space: Proxemic Behavior of Children 6
through 16," *Human Ecology*, Vol. 2, No. 3, July, 1974, pp. 177-189.

KEY WORDS:

Man	Setting	Relational Concepts	Outcomes
Children	*Schools*	*Proxemics*	*Social Behavior*

     This study is an investigation of the effects of age and sex on the
interrelated measures of distance and bodily orientation. The authors
hypothesize that boys and girls exhibit similar proxemic behavior during
their early years, but that males, in later years, would maintain
greater distances and axial body orientations.

     Four hundred and twenty-four white school children between the ages
of 6 and 16 living in a suburb of Lansing, Michigan, were the subjects
of this particular study. After an initial screening process to assure
consistency and minimize statistical error, 196 same-sex pairings were
chosen to measure social distance and axial orientation. Distance was
measured by a 14 point proxemics scale while orientation was measured by
a 25 point scale from 0 meaning a pair facing one another to 24 meaning
a pair standing back to back. During a prearranged 30 second discussion
between each pair approximately eight observations of axis and distance
were measured by one of six judges. Inter-judge reliability for both
measures was maintained at 90 percent or greater during the two month
test.

     The major conclusions of the study were that in support of the ori-
ginal hypothesis, males exhibit a gradually increasing need for greater
personal space and that sex differences in young children do not exist.
Specifically, it was found that adult proxemic attitudes are established
by age 12 with females requiring 17 inches of interaction space as
opposed to 27 inches for males. It was also shown that while male body
orientation remains fairly constant throughout maturation, the axial
alignment of females becomes more intimate and direct as age increases.
The authors suggested that social attitudes concerning manliness and
competitive feelings might explain the increasing personal space
requirements of young male adolescents.

"Crowding and the Role of Interpersonal Distance Preference,"
Sociometry, Vol. 40, No. 3, September, 1977, pp. 271-282.

KEY WORDS:

Man	Setting	Relational Concepts	Outcomes
Female Students	*Laboratory*	*Crowding*	*Stress*

     An experiment was performed at Rutgers University to look into the
role of interpersonal distance preference as a mediator of short-term

crowding. The hypothesis was that those subjects preferring "far" interpersonal distance would experience greater stress in a crowded environment than subjects with closer interpersonal distance preference. During the crowding experience (subjects sat with knees and shoulders touching other subjects seated in a small room), subjects preferring "far" interpersonal distances were found to be most stressed, as measured by electrodermal responses and by lower performance levels on creativity tasks administered immediately after the crowding experience.

The subjects involved were 32 female undergraduates enrolled at Rutgers University and the small-sized room used in the experiment was only 4 x 2.5 ft., constructed within a larger existing room with an 8.5 ft. ceiling.

The results indicated that for subjects who feel comfortable interacting with others at close distances, this intense crowding did not increase stress levels. The implications are that the negative effects of a stressful environment are perceived as even worse if that individual feels a lack of control.

This study also indicated that crowding, regardless of interpersonal distance preference, has an effect on creativity due to decreased frustration tolerance. Exposure to the crowding experience was very brief (30 minutes for each group of 4) and, therefore, has no bearing on problems where crowding is on-going, such as work situations. It is more relevant to mass-transit settings but also differs in the fact that no danger is presented, heat and noise are no problem, no members of the opposite sex were present, and no control was given to subjects as far as choice of entry.

The subjects expressed negative feelings concerning the crowded environment but did not generalize those feelings to include their fellow group members.

ALLEN, M.A. AND G.J. FISCHER   3

"Ambient Temperature Effects on Paired Associate Learning," *Ergonomics,* Vol. 21, No. 2, 1978, pp. 95-101.

KEY WORDS:

Man	Setting	Relational Concepts	Outcomes
Male students	*Laboratory*	*Thermal comfort*	*Learning performance*

Past studies have shown that temperature variations in a classroom can affect the learning abilities of the subject children. The authors believed that while temperature variations over a constant relative humidity will indeed affect the performance of subjects, if variations are matched with a commensurate change in humidity and a relative comfort zone is maintained, learning will not be impaired.

Two experiments were conducted in which each of the stated hypotheses were tested. In the first experiment, 65 male university students were given pair-associate (P-A) memory and retention tests over a range

of temperatures and humidity readings of 52°F and 57% to 92°F and 21%.
A standardized memory test was administered to each group of subjects at
each comfort range and skin temperatures were monitored to record physiological changes in each subject. The second test involved the same
learning exercises administered to 84 male students, this time with both
wet and dry bulb temperatures being varied over a constant relative
humidity level. This approximated a continuous comfort reading over
various temperature settings.

In the first experiment, the data showed a high correlation between
high learning abilities and a temperature setting of 72°F. In temperatures above and below this reading, learning abilities were measurably
diminished. When a comfort coefficient is held constant over a range of
temperatures, as in the method used in the second experiment, test
scores held constant at temperatures between 52° and 82°F. Even though
skin temperatures varied 2° with every 5°F shift in ambient measures in
both experiments, the authors concluded that comfort rather than absolute thermal readings is the key variable in determining resultant
learning behavior.

ALLEN, T.J. and P.G. GERSTBERGER   4

"A Field Experiment to Improve Communications in a Product Engineering
Department: The Nonterritorial Office," *Human Factors,* Vol. 15, No. 5,
October, 1973, pp. 487-498.

KEY WORDS:

Man	Setting	Relational Concepts	Outcomes
Office workers	*Environmental arrangement*	*Territoriality*	*Communication/ performance*

Open office plans have been used as a means to improve the communication among individuals working in a common office environment. This
study tests the hypothesis that the use of open office plans does in
fact have a positive effect on interpersonal communication and that
workers can function in a nonterritorial type of work setting.

A product engineering section of a large manufacturing concern was
used as the experimental environment. Twenty-four product engineers who
had previously worked in private offices were relocated to an open
office area that contained no personal equipment or furnishings. All
tables and work materials were shared by all members of the group and no
partitions or barriers were constructed within the work area. A control
group of a similar company's engineering department was used to measure
the effects of the traditional work environment. Both groups were
administered questionnaires concerning levels of noise, degrees of privacy, and ease of communication two months prior to, three months
after, and five months after the change from traditional private offices
to the open plan concept. The experimental group was also sampled once
each week over a period of one year to determine patterns of intra and
inter-departmental communications and seating arrangements within the
work space.

The results of the experiment reinforced the hypothesis that inter-personal communication would increase in the open office scheme. Although perceived noise levels increased in the open plan, overall worker satisfaction was 15 to 25 percent higher than that of the traditional office layout. Less that 10 percent of any given worker's time was spent at a single work table which led the authors to conclude that territoriality was not an essential element of a work environment. Overall production levels, however, were not increased within the experimental work area and communications outside the plant were unaffected by the facility change.

AMICK, D.J. AND F.J. KVIZ   5

"Social Alienation in Public Housing: The Effects of Density and Building Types," *Ekistics*, Vol. 39, No. 231, February, 1975, pp. 118-120.

KEY WORDS:

Man	Setting	Relational Concepts	Outcomes
Human	*Public Housing*	*Density*	*Social alienation*

Past studies have established the correlation between feelings of social alienation and high housing density. The author of this study suggests that density should be measured not only in terms of housing size per person but should also include a measure of the population cover on a given building site (high-rise vs. low-rise). He hypothesizes that high-rise dwellers will suffer more from alienation than those who occupy low-rise apartments at comparable dwelling densities.

Nine hundred and fifteen occupants of six Chicago area low-income housing developments (368 low-rise and 547 high-rise dwellers) were interviewed to determine their feelings of social isolation and power-lessness. A series of four questions were asked concerning their response to specific social situations and a scale of alienation was developed from these responses. Each of the six developments were given an Index of Interaction Restriction which measured the family dwellings per acre to the percent of total site acreage used for buildings. This was an index which measured density in terms of high-rise vs. low-rise factors.

The survey showed a high correlation between social alienation and the occupancy of a high-rise apartment. The high-rise, with an index of 364, was shown to have a 41.5% degree of alienation, while the low-rise dweller, with an index of 84, had an alienation rating of only 26.6%. The actual population density per dwelling, however, was quite similar (5.85% versus 4.19%), which lead the author to conclude that the high relationship between an interaction index and alienation ($r = .89$) is much more significant than pure density figures. In the comparison of inner-urban versus suburban low-rise dwellers, a decrease in alienation (44.6% versus 34.8%) in the rural setting was found, even though an increase in density from 2.54% to 4.77% was recorded.

"Human Perception of Humidity Under Four Controlled Conditions,"
Archives of Environmental Health, Vol. 26, No. 1, January, 1973, pp. 22-
27.

KEY WORDS:

Man	Setting	Relational Concepts	Outcomes
Male students	*Thermal environ- ment*	*Thermal comfort*	*Humidity sensations*

The sensation of thermal comfort which is experienced by an individual is a function of many factors, such as temperature, humidity, air
pressure, clothing and activity level. This study was designed to
determine how the human perception of humidity changed when all of the
other variables were held constant while relative humidity was raised or
lowered.

During the period from September 22 to October 23, 1970, tests were
conducted in the climate chamber at the University of Aarhus in Denmark.
Forty-eight healthy Danish male medical students took part in the experiment. Three subjects were placed in a special uniform with an insulation value of 0.7 clo. They were seated at small tables and were only
allowed to read the paper or perform other sedentary type activities.
Temperature in the chamber was held constant at the $23°+ -0.3°$ C mark,
regardless of the humidity level.

The actual testing consisted of eight hours. In all cases, relative humidity (RH) was set at 70% for the first two hours. During the
remainder of the experiment different sequences of 50%, 30%, and 10% RH
were utilized. Every half hour subjects were given ballots and asked to
vote regarding perceived temperature and humidity sensations for specific body surfaces such as the eyes, nose, lips, mouth, pharynx, hands,
face, and hair. The Bedford seven-point rating scale was used, with all
ballots concealed from the others in the room. No discussion was
allowed.

The results revealed "no change in the perception of humidity on
lowering it from 70% to either 50%, 30%, or 10%, nor on returning it to
the initial level of 70%." However, although the changes in relative
humidity were not recognized as such, highly significant changes in the
sensation of temperature were reported. The authors concluded that "the
human perception of a change in water vapor pressure in either direction
is negligible in comparison to the thermal effects aroused by this
change."

"Indoor Air Pollution Due to Chipboard Use as a Construction Material."
Atmospheric Environment, Vol. 9, 1975, pp. 1121-1127.

KEY WORDS:

Man	Setting	Relational Concepts	Outcomes
Human	*Indoors*	*Pollution/build- ing material*	*Contamination*

Materials which emit harmful vapors are normally regarded as a problem only in confined spaces designed for long-term stay, e.g. nuclear submarines or manned spacecraft. The problem, however, is of general hygenic concern since a multitude of air contaminants are being generated within the home environment from various building materials, fixtures, furniture, textiles and other household products. This paper examines the emanation of formaldehyde from chipboard used as a building material.

The use of chipboard has increased rapidly during the last decade. In the home the material is used for partitions, ceilings and floors, for wardrobes, and furniture. Chipboard consists of woodshavings held together by glue, usually an urea-formaldehyde glue. During the adhesion process most of the formaldehyde is released, but some free formaldehyde remains in the board, which in the course of time is released and partly replaced by formaldehyde regenerated from the polymerized urea-formaldehyde glue.

Field measurements were conducted from Feburary to September, 1973, and included 25 rooms in 23 dwellings (19 houses and 4 flats), all situated in suburban areas in Jutland, Denmark.

The equipment used for measuring the level of formaldehyde was placed in the room before the measurements. Doors and windows were closed, and only the natural ventilation through leaks and cracks remained. The radiators were set for a room temperature of 20-25° C, and the measurements were carried out in rooms without furniture. As a control, measurements were also conducted in the climate chamber at the Institute of Hygiene, Aarhus, Denmark. Fourteen boards from the normal production line of a Danish factory were placed in the room.

Formaldehyde may affect the human body by absorption through the skin, inhalation or by oral intake, of which the last is of no importance in the home environment. In dermatology, toxic or allergic skin infections caused by formaldehyde are often encountered. Occupational allergy to formaldehyde is well known, but so far it has not been possible to state a threshold for development of skin disease. The adverse health effects of formaldehyde are due to its great affinity with water and its protein-denaturing property. The most important effect of low concentrations of formaldehyde is irritation of the mucus membranes in the nose and in the upper airways.

In the study the threshold limit values for occupational exposure and the limit values for continuous outdoor exposure to formaldehyde in the U.S.A., West Germany and U.S.S.R., was exceeded in two dwellings, and only two dwellings had lower concentrations than the American value for outdoor exposure. These dwellings were 3-5 years old and had a very low content of chipboard. The study concluded that the use of chipboard in its present form, may result in higher indoor formaldehyde concentrations than permitted for continuous outdoor exposure. The authors further suggested that the rather extensive system of air quality standards and control programs for workroom air and for community air should be supplemented by air quality standards and control procedures for indoor air in the home.

"The Place of Architectural Factors in Behavioral Theories of Privacy,"
The Journal of Social Issues, Vol. 33, No. 3, 1977, pp. 116-137.

KEY WORDS:

Man	Setting	Relational Concepts	Outcomes
Human	*Built environment*	*Privacy*	*Environmental analysis*

This article proposed a model of the environment for conceptualizing privacy and other forms of interpersonal behavior. The bases for
this model were the physical properties of the architectural environment, the behaviorally relevant attributes of the environment, and the
role of the physical environment in the presentation of information
about the self and in the experience of privacy. Privacy was considered
a relational characteristic or attribute. Archea maintained that "each
person is the center of a dynamic field of information about surrounding
events and activities, to which his or her behavior is a continuous
adjustment." And that, "the arrangement of the physical environment
regulates the distribution of the information upon which all interpersonal behavior depends."

One's orientation in space limits the amount of information about
his or her immediate surroundings which can be acquired at a single
moment. The ratio between the maximum amount of a setting that can be
observed while oriented in a single direction and the total amount that
can be observed using all possible orientations from a given location
determines the efficiency of visual access. According to the model, the
physical environment is a relatively stable assembly of walls, doors,
corners, and other regulations of the flow of information. The physical
arrangement of a given setting remains substantially unchanged from one
situation to the next. The dynamic influence of the environment or
interpersonal behavior stems from the ways in which it is used. By
selecting or changing one's location or orientation, one establishes a
potential for obtaining and conveying behavioral relevant information.

"Subjective Annoyance from Noise Compared with Some Directly Measurable
Effects," *Archives of Environmental Health,* July/August, 1978, pp.
159-165.

KEY WORDS:

Man	Setting	Relational Concepts	Outcomes
Male students	*Sonic environment*	*Noise*	*Task performance*

Much debate surrounds the potential consequences that noise implies
for an individual's mental and physiological state. In response to that
debate, this study compared subjective reports of annoyance to noise,
with performance on an arithmetic test. In addition, stress hormone

levels were measured in order to gain an understanding of the physiological reactions of exposure to noise.

One hundred male university students (23-30 years of age) were randomly separated into four groups of 25. Each group went through four 15 minute experimental sessions in different sequences: (1) performance test with noise, (2) performance test without noise, (3) reading a paper with noise, and (4) reading a paper without noise. Responses regarding annoyance were recorded according to a verbal category scale.

Prior to each day's experiments, subjects worked practice problems similar to the ones done under test conditions to make sure that the task was understood by each individual. Physiological responses were gathered in the form of urine samples taken before and after the test sessions. Freeway traffic sounds provided the background noise. The sound level was moderate compared to many earlier laboratory studies of noise, but still greater than is common in residential areas.

The results revealed that "the excretion of stress hormones and the performance efficiency in sessions of exposure to noise as compared with sessions of no noise, no effect of noise can be discerned." The same result was obtained when the noise versus no-noise relaxation situations were compared.

When the subjects were grouped according to their stated feelings of annoyance, however, it was discovered that exposure to noise affects performance more negatively in the "more annoyed" individuals. Thus, the authors hypothesize that "the tendency to be annoyed by noise is associated with impaired performance in a generally stressful situation. In addition, feelings of annoyance may partly be caused indirectly by the subject's awareness of his performance in the noisy environment."

ASHTON, R.   10

"The Effects of the Environment Upon State Cycles in the Human Newborn," *Journal of Experimental Child Psychology,* Vol. 12, No. 1, August, 1971, pp. 1-9.

KEY WORDS:

Man	Setting	Relational Concepts	Outcomes
Infants	*Hospital labora-tory*	*Noise/illumina-tion*	*Comparative response levels*

This study assessed the effects of different levels of ambient noise and illumination in a controlled hospital environment upon both the ongoing state of newborn infants and upon related heart and respiration rates. It was hypothesized that organic states were amenable to experimental manipulation and that it was possible to assess such effects. Also, these physiological parameters could be monitored over long periods of time. The effects of a high ambient noise level and a low level of illumination were compared to results obtained from a control group of babies subjected to standard environmental conditions. The author maintained that the ambient illumination level upon neonatal behavior has never been examined properly, although the effects of noise upon activity have been shown to influence infact behavior.

Twenty-two full-term, healthy neonates were used as the control group. Ten neonates were used in each of the two experimental groups. The mean ages of the control group was 72.80 hours; mean of subjects in the dim-light conditin was 68.45 hours; and mean age was 79.50 hours for subjects in the ambient noise condition.

The testing procedure consisted of the infants being placed in a plexigalss crib in a hospital laboratory with one of the experimental conditions operating. Infants were randomly assigned in one of three conditions (high ambient noise, low illumination level, standard conditions). The various parameters (heart rate, respiratory pattern, state, activity, crying, and eyes open or closed) were measured. The recordings taken from each infant varied in length, but all exceeded two hours.

As compared to the control group of subjects only the dimmer light condition had any effect. It was found to stabilize the respiration rate per minute across subjects irrespective of state and to significantly lower the mean respiration rate according to measures determining either Alertness or Active Sleep. The increased level of ambient noise had no such effect. Neither environmental change significantly affected the cycling of the various physiological states, nor had an effect upon infants' mean heart rates per minute. It was concluded that the infant, like the adult, "is more susceptible to changes in the environment than to the ongoing, unaltered nature of that environment."

ANUOLA, S., R. NYKYRI AND H. RUSKO   11

"Strain of Employees in the Machine Industry in Finland," _Ergonomics_, Vol. 21, No. 7, 1978, pp. 509-519.

KEY WORDS:

Man	Setting	Relational Concepts	Outcomes
Workers	_Factory_	_Occupation_	_Strain_

Strain in the work place is often measured by the physiological effects on workers in specific types of occupations and job settings. The authors of this study propose to show the relationships between different environments of the machine industry and the resulting expenditure of body energy.

One hundred-ninety men and forty-seven women employees of three Finnish machine factories were used as subjects for this experiment. Twenty-six common occupations in the factories were selected as typical representations of heavy industry work and three clerical occupations were selected as a control group. The subjects were monitored throughout the spring of 1973 in terms of heart rate, oxygen consumption, and caloric expenditure through the use of physiological device readings taken at specified times during the workday. The most important stress measurement was deemed the relative heart rate, defined as a percentage of working heart rate divided by a subject's maximal heart rate.

It was found that strain in male workers was greatest in the early
stages of production work, especially in occupations such as foundry and
forging environments. The relative oxygen consumption of semiskilled
men and women was very similar in all age groups, particularly in the
older brackets. "Although the absolute strain at work in women in
general was slighter than men, the values of energy expenditure at work
tended to be a little higher in the oldest female age group than in the
oldest group of skilled men." It was also found that, although absolute
strain and relative heart rate in skilled and semiskilled men in the
lower age group was the same, semiskilled men 45 years and older had
significantly higher strain values. The authors concluded that the primary variable in the measurement of work strain was age and that work
conditions should be considered over a long-term period in order to compensate for changing physical capacities of workers in older age
brackets.

BAIRD, J., B. CASSIDY, AND J. KURR   12

"Room Preference as a Function of Architectural Features and User
Activities," _Journal of Applied Psychology,_ Vol. 63, No. 6, December,
1978, pp. 719-727.

KEY WORDS:

Man	Setting	Relational Concepts	Outcomes
Students	Rooms	Architectural elements	Preferences

     In this study of architectural room preferences, the authors
hypothesized that the types of preferences were a function of perceived
architectural features and the activities that occur within a series of
rooms. A group of undergraduate psychology students comprised that subject group. Subjects were made aware of a series of activities and spatial changes occurring within a set of simulated room environments and
in two of the three sub-experiments the architectural stimulus was
represented by 20 scale models. A brief literature review preceeded
presentations of the studies. This review discussed the value of
exploring the empirical link that exists between environmental design
and user behavior and attitudes.

     Three experiments were conducted on preferences for simulated rooms
as a function of variations in architectural features (ceiling height,
ceiling slope, and wall angle) and variations in presumed user activities (talking, listening, reading, dancing, dining, and no activity
specified). In Experiments 1, 2, and 3 the sample sizes (N) were 30,
30, and 13, respectively. Subjects rated the desirability of simulated
rooms (and semi-fixed features of rooms) on a response scale in all
three experiments that ranged from -10 to +10. Different subjects were
used for each experiment, although they were all enrolled in the same
Introductory Psychology course. Descriptive statistics were utilized in
all three experiments. Mean preferences were analyzed as a function of
the logarithm of each of the three major types of room variations and
analyses of variance were performed.

In Experiment 1, the architectural feature was varied while user activity was fixed. In Experiment 2, the user activities were varied for a constant architectural feature. Ceiling heights were the focus, and each individual rated each model room for the six activities (mentioned above) while imagining he or she was actually in the room. The results for the talking and dining activities were virtually identical.

In Experiment 3, an adjustable ceiling was constructed that would be fixed at different heights and allowed subjects to walk beneath and around it in order to gain a more realistic view of the desirability of the different heights. The major question was whether the preference ratings given under these circumstances would agree with those given for the model rooms. It was found that subjects' preferences were more sensitive to changes in the adjustable ceiling than they were to changes in the models. The general characteristics of the preference functions were the same when the architectural stimulus was a model at one inch = one foot or when a small ceiling whose height could be varied from 6 to 15 feet was rated. In all instances the data on ceiling height were modeled as a quadratic equation.

The authors concluded that this study had both theoretical and applied implications for architects and other design professionals. In general, subjects preferred ceilings that were higher than those normally encountered in the built environment, although the authors felt that development of more sophisticated preference models must await further studies with more numerous and varied subject populations.

BAKEMAN, R. AND R. HELMREICH   13

"Cohesiveness and Performance: Covariation and Causality in an Undersea Environment," *Journal of Experimental Social Psychology*, Vol. 11, September, 1975, pp. 478-489.

KEY WORDS:

Man	Setting	Relational Concepts	Outcomes
Research teams	Undersea habitat	Group cohesiveness; performance	Performance influences cohesiveness

The relationship between group cohesiveness and performance was investigated using data from the observation of persons in an undersea habitat. Individual performance (productivity) was defined as the percentage of time each aquanaut within a team was observed via unobtrusive measures as being engaged in work activity. Performance of each of 8 teams was viewed as the mean performance of its members. Each team of individuals was comprised of trained marine scientists who were selected for participation in an underwater research study that allowed freedom of research topics for each scientist. Subsequent to the dive, 22 reports were prepared, based on work completed while in the habitat. These reports were rated for content and scope by a team of judges. Group cohesiveness was seen as a member's attraction to the group. Attraction to the group was measured by conversational behavior in leisure time since work activities required certain "business oriented" conversation, which were usually conducted during routine daily tasks.

Individual gregariousness was defined as the percentage of time an aquanaut was observed in leisure time conversation relative to when such behavior was possible. The authors noted that these definitions were strictly operational and were not be be misinterpreted in a broader context.

The research setting was Project Tektite 2, a series of saturation dives conducted in the U.S. Virgin Islands in 1970. Ten teams of aquanauts spent a total of 182 days living in an underwater habitat and working on the surrounding reef. Forty-eight subjects were observed by a psychological research team 24 hours a day.

Psychological data were collected by surface-stationed observers who continuously monitored in-habitat behavior via closed circuit television. Audio data were collected via microphones. The observers sampled ongoing behaviors and activities every 6 minutes and systematically coded all data. These measures were coupled with each team's reseach reports in order to test hypotheses. A series of regression analyses measured in interactive effects between major variables.

Results indicated a high correlation between an index of leisure time cohesion and work performance. This accounted for 42% of performance variance. Although many prior studies assumed that cohensiveness caused performance, the authors stated the view that in many contexts the dominant direction of causality may be from performance to cohesiveness instead. By their nature, laboratory experimental studies impose a causal direction and were viewed by the authors as inappropriate in answering the type of questions posed in this study. It was concluded that in the context of this environmental setting, cohesiveness was not an important determinant of performance. In contrast, positive, productive team performance in work activities caused group cohesiveness.

BALDASSARE, M.   14

"The Effects of Density on Social Behavior and Attitudes," *American Behavioral Scientist,* Vol. 18, No. 6, July/August, 1975, pp. 815-825.

KEY WORDS:

Man	Setting	Relational Concepts	Outcomes
Urban dwellers	Densely populated city area	Densely populated environment/social behavior	Low neighboring

In this article the author addresses the question, What are the effects of dense urban environments on the social relations of individuals? The study examines some of the social-psychological effects of dense urban environments upon individuals. In doing so, it assumes two characteristics of high-density urban individuals. First, that crowded city dwellers develop certain social adaptations to cope with the "psychic overstimulation" of high-density living. Second, that crowding will cause a decrease in social behavior and related personality attributes.

The data for the study was obtained from the Detroit Area Study, which was conducted in 1965-1966. Four socio-behavioral questions were included in the analysis. These questions were concerned with the respondents' number of friends, days per week spent with friends, number of organizational affiliations, and how well individuals knew their neighbors.

The study concluded that dense urban neighborhoods may be related to low neighboring among residents. However, considering the data, the author was only able to conclude that density has, at most, only local effects on social relations.

Past studies offered only speculative hypotheses on high-density urban areas and their effects on individual social behavior. Four possible explanations given in the study were: 1) urbanites interact less with their neighbors and are not spatially constrained to choose friends from the immediate vicinity, 2) less neighborly people may be attracted to these higher-density areas or may have no choice but to live in these areas, 3) urbanites may perceive themselves less neighborly simply because they have so many neighbors, and 4) crowded urbanites may withdraw from local contacts (i.e., neighbors) as a specific adaptation to high levels of stimulation.

BALLANTYNE, E.R. AND J.W. SPENCER   15

"Climate and Comfort in a Humid Tropical Area," *Build International,* July/August, 1972, Vol. 5, No. 4, pp. 214-219.

KEY WORDS:

Man	Setting	Relational Concepts	Outcomes
Tropical householders	*Port Moresby, New Guinea*	*Thermal factors, building materials*	*Thermal comfort*

The effect of a humid, tropical environment on the indoor temperatures at different times of the year poses various design difficulties for architects building in those regions. However, little is known regarding the influence of the various climatic parameters in existence. Port Moresby, the site for this extensive tropical climate study, is situated on the southern coast of the Territory of Papua and New Guinea, Located at a latitude of 9 1/2° S, Port Moresby has remarkably constant weather patterns compared to those cities located in more temperate regions. Hence, the aim of this study was to calculate indoor temperatures at different times of the year in houses of different constructions and orientations, in the hope that several of the findings might be generally applicable to humid tropical localities.

The authors gathered climatic data over a period of a year for temperature, relative humidity, rain, wind velocity and direction, and solar radiation. At the same time a survey was undertaken to ascertain the temperatures preferred by acclimatized residents in Port Moresby. The survey consisted of asking the occupants of sixteen houses to report the thermal sensation which they were experiencing at various points in time. A seven-point scale was used to measure thermal comfort: much too cool; too cool; cool; neutral; warm; too warm; and much too warm. Temperatures in the houses were continuously monitored and recorded on thermographs. Clothing was held fairly constant throughout the year. In all, a total of 3600 assessment cards for 34 subjects were gathered and analyzed.

Results from the analysis revealed that the maximum number of subjects felt neutral (no desire to be cooler or warmer) at 25.4° C. The range of temperatures for a neutral assessment generally fell between 23.4° C and 27.6° C, and for a warm assessment between 27.6° C and 31.8° C. In addition, it was noted that, although higher temperatures became more tolerable as air movement increased, the neutral decreased at the same time.

Housing type proved to be a factor. The occupants of concrete houses had a significantly higher preferred temperature compared with occupants of weatherboard and asbestos cement houses, while the occupants of aluminum houses had a significantly lower preferred temperature. Furthermore, calculations of indoor temperatures showed that an east-facing room in June was hotter than a west-facing room, with a north or south-facing room in June being the best of all the orientations examined. Thus, the authors concluded that "buildings should be sited with their long axes running east and west with necessary glazing in their north and south walls. At the latitude of Port Moresby it would be appropriate to have main rooms facing south, with a passage on the north side to reduce unavoidable heat gains during May, June, and July."

BALTES, M.M. AND S.C. HAYWARD   16

"Application and Evaluation of Strategies to Reduce Pollution: Behavioral Control of Littering in a Football Stadium," _Journal of Applied Psychology,_ Vol. 61, No. 4, August, 1976, pp. 501-506.

KEY WORDS:

Man	Setting	Relational Concepts	Outcomes
Spectators	Football stadium	Littering control	Nonlittering behaviors

It is assumed that littering and nonlittering behaviors can be conceptualized as operants, since they are acquired, maintained, and modified by the same principle as other learned behaviors. The littering event is viewed here as a process that revolves around sets of priorities where both positive and negative consequences can be experienced. In this study an attempt was made to modify adult littering behavior in the football stadium at Pennsylvania State University.

Four different treatment strategies were used: 1) an operant contingency in the form of a positive reinforcement procedures, 2) a positive prompting strategy, 3) a negative prompting strategy, and 4) a litterbag - only condition.

The sample group was comprised of 1200-1700 nonuniversity season ticket holders in each of six seating sections in the stadium. The testing conditions were applied twice in two consecutive games. A between-subjects design was chosen as the experimental strategy. This required random assignment of subjects to treatment and control groups allowed for the comparison of treatment effects confounded with sequence effects. Each subject was provided with a clear plastic bag with one of three types of written instructions attached (positive reinforcement, positive promptings, or negative prompting). The fourth group (litter-bag-only condition) received no information and the no-treatment group received neither litterbags nor instructions.

The dependent variables were: a) the weight of improperly deposited litter per row, and b) returned litter weight per treatment group. Other methods were used to weigh the remaining litter.

The findings revealed a highly significant main effect of treatment, which was responsible for 45% reduction in the amount of litter in the treated sections of the football stadium. There were no significant differential effects between the different treatment strategies. Thus, all four treatments significantly reduced littering when compared with the no-treatment conditions.

The authors concluded that these findings provided a clear illustration that even in "legitimate" littering situations (such as sporting events), appropriate treatment strategies can effectively modify persons' behaviors, resulting in nonlittering behavior. Also, they speculated that among the many advantages possible, the design of litterbags could aid in encouraging nonlittering behaviors, and economic savings can be incurred if the lessened amount of litter required less time and personnel in the cleanup process.

BANZIGER, G. AND K. OWENS   17

"Geographical Variables and Behavior: II. Weather Factors as Predictors of Local Social Indicators of Maladaptation in Two Nonurban Areas," *Psychological Reports,* Vol. 43, No. 2, October, 1978, pp. 427-434.

KEY WORDS:

Man	Setting	Relational Concepts	Outcomes
Mental patients	*Institution; two nonurban areas*	*Maladaptation; social indicators, geographical variables*	*Weather factor predictors*

The relative predictive strengths of eight weather factors in the environment were examined using a separate dependent variables monthly figure for: a) community mental health intake; b) welfare caseload; c) calls to a telephone hotline; d) medical patient caseload; e) felony arrests; f) juvenile complaints; g) drunk driving arrests; and h) mortality rates. Data was collected from two nonurban areas of Ohio. The influence of seasonal changes on the psychological status of humans has long been a subject of subjective observation. It has often been assumed that inclement weather, i.e., snow, rain, extreme cold, is responsible for a negative disposition in people and that the control

and protection of people from these conditions would make them well
adjusted, as Skinner has suggested in his fictional Utopia.

Prior researchers have studied the relationship between climate
conditions and mental health in institutional settings. Suicide rates
have been linked to weather conditions. A number of relevant studies
were reviewed by the authors.

In order to determine the relative subjective importance of each of
the eight chosen weather factors, three independent samples were asked
to rate on a 5-point scale each factor according to its contribution to
the subjective discomfort caused by the weather. Eight weather factors
were rated. Monthly summaries of weather data for both locations studied were obtained for a 39-month period. The two locations had comparable community mental health centers, facilities for out-patient
counseling, and preventive mental health services.

Z-score transformations of subjective discomfort of weather factors
as indicated by the three independent samples were analyzed with a stepwise multiple regression. With the exception of hotline calls, each of
the social indicators in the two localities was significantly predicted
by a different weather factor, and the weather factors, taken together,
accounted for about 10% of the variance of each social indicators. For
each geographical area, combined weather factors accounted for no more
than 30% of the variance of any local social indicator. The authors
concluded that problems concerning the effects of weather factors are
not unfounded, and that a balanced approach to behavioral effects of
geophysical variables must be achieved.

BATCHELOR, J.P. AND G.R. GOETHALS   18

"Spatial Arrangements in Freely Formed Groups," *Sociometry*, Vol. 35, No.
2, June, 1972, pp. 270-279.

KEY WORDS:

Man	Setting	Relational Concepts	Outcomes
Students	Laboratory	Collective decisions	Spatial arrange-ments

In an investigation of the spatial ecology of groups, it was
hypothesized that the way the group functioned would affect the spatial
arrangements of the members. Senior high students from a regional rural
school in Massachusetts volunteered to participate in the experiment and
they were divided into groups of eight, 4 males and 4 females. An empty
cafeteria room was used and the experiment was presented as research on
the differences between group decisions and individual decisions. The
subjects were told to read a case history concerning a delinquent boy
and seven different ways of handling him. Some groups were told to make
individual decisions about the case without consulting anyone else while
other groups were told to discuss the case and try to reach a collective
decision. All groups were then to enter the room and "take a chair and
sit any place."

In a group of 8 there are 28 interpersonal distances for each group. The total of these distances gives an indication of the size of the group, i.e., the higher the score, the more the group was dispersed, giving it greater size. A second measure, distance to the nearest person, was arrived at by averaging the distance to the nearest person for each person in the group.

The results showed that the independent variable, the instructions to decide individually or collectively, had significant effects on the spatial arrangements of the members of the groups. When the subjects had to work as a group, the spatial arrangements of chairs were almost uniform and the conformity of interpersonal distance was great within these groups. However, in the individual decision groups there were wide variations in the size of the groups. Some groups were as small as collective decision groups and some used the whole room, but the amount of separation within a given group tended to be constant.

In these mixed-sex groups the person sitting closest to a given subject was more often the same sex than would be expected by chance. It is strongly suggested by these results that norms exist in the use of space.

BAUM, A. AND S. VALINS   19

"Architecture, Social Interaction and Crowding," *Transactions of the New York Academy of Sciences,* Vol. 36, No. 8, December, 1974, pp. 793-799.

KEY WORDS:

Man	Setting	Relational Concepts	Outcomes
Students	*Laboratory*	*Crowding/ residential environment*	*Values-issues problems/experimental data, use of*

At the State University of New York, Stonybrook, laboratory experiments were developed to evaluate the effects of residential environments on the behavior and mood of students. In this study, which took two years to complete, groups of freshmen who lived in two very different dorm situations were involved in four experiments. The subjects lived in either standard corridor-type dormitories, 17 rooms to a floor with a common lounge and bathroom at one end, or in dorms built with a basic suite design of 2 or 3 bedrooms with a shared bath and lounge. The hypothesis to be evaluated was that in interacting with 33 other people, as opposed to 4-5 people in the suite-type dorms, more social stress would be produced.

The first experiment involved the use of miniature rooms and figures of people. Each subject (100 freshmen of both sexes) was asked to place miniature people in three different model rooms until they felt each room was crowded. The model rooms chosen were a small bedroom, a lounge area and a library reference room. Data indicated that the corridor-design dorm persons placed significantly fewer people in the bedroom and lounge area but more in the reference room. This was interpreted to mean that these subjects disliked having to interact with a large number of persons.

In another experiment, designed to test interaction, 64 male and female freshmen were used from both types of dormitories. A subject was brought into a room where another person of the same sex was already seated in one of seven chairs. The subject was asked to take a seat and such factors as seat position, visual movements and talking were noted through two-way mirror windows for 5 minutes. Corridor-type dorm subjects sat further away from the other person, avoided eye contact and said they felt more uncomfortable than residents of suites.

The last two experiments examined the hypothesis that an environment in which persons are actively avoiding social interaction is one in which competitive orientations may develop. Competition, in this case, represents less intense involvement with people whereas cooperation may indicate a deeper commitment to an individual or group.

Forty freshmen of both sexes were selected from each type of dormitory and assigned to cooperate on a task with others or to do a task alone. Data collected indicated that suite residents did best at tasks alone while residents of corridor-type dorms did best when co-acting with others, since they interpreted it as a competitive situation.

In the last experiment, the cooperative and competitive orientations were more explicitly aroused with 64 different freshmen students by adding the incentive of pay. Data showed that the suite-resident subjects did best alone or cooperating while the corridor-type dorm residents performed best when competing with others which actually means avoiding interaction. The corridor-type dorm resident feels crowded, tolerates fewer people in social spheres and avoids social interaction.

The authors conclude that it is possible to base design of buildings on experimental data but the value-issues problem must be confronted and a judgement made on which behaviors are considered healthy and which are not.

BAUM, A., M. RIESS, AND J. O'HARA   20

"Architectural Variants of Reaction to Spatial Invasion," *Environment and Behavior*, Vol. 6, No. 1, March, 1974, pp. 91-100.

KEY WORDS:

Man	Setting	Relational Concepts	Outcomes
Students	*Drinking fountain*	*Personal space*	*Invasion*

Previous personal space studies dealing with the considerations of spatial invasion involved observation of the reactions of a passive subject to invasion by an experimenter. This study acknowledged that there is a limit beyond which individuals may not approach each other without experiencing some discomfort. It was hypothesized that the presence of architectural constructs or physical barriers, which reduced the impact of one individual on another, would moderate reaction to the invasion of personal space.

The setting for the study was an academic building on a college campus. Within this structure, two different types of water fountains were studied. Subjects were defined as students passing in the hallways during experimental sessions who ventured within five feet of the water foundation as they walked through either one of the two corridors where the fountains were located. A total of 310 subjects were sampled. A 2 x 3 design was utilized to consider three proxemic conditions crossed with the two architectural settings. Proximity was varied by positioning a confederate against the wall near the fountains at two spatial intervals. Data were also collected for a control condition at each fountain where the confederate was not present.

This examination of personal space behavior which placed the burden of activity on the subject yielded data generally confirming the expectations of the authors. Subjects drank more often from a screened water fountain than from a water fountain without such barriers. Also, sex differences were found between the intruder's two proxemic conditions (intervals) regarding the total number of subjects drinking. The close presence of the intruder intimidated subjects and activity discouraged them from using the water fountains.

It was concluded that physical factors can alter spatial perception and related spatial behaviors. Architectural design and the location of interior amenities were found to influence the manner in which an individual moves through space. The need to incorporate personal space data within the architectural design process was stressed by the authors.

BECHTEL, R.B. AND A. GONZALEZ   21

"Comparison of Treatment Environments Among Some Peruvian and North American Mental Hospitals," *Archives of General Psychiatry*, Vol. 25, No. 1, July, 1971, pp. 64-68.

KEY WORDS:

Man	Setting	Relational Concepts	Outcomes
Psychiatric patients	*Psychiatric hospitals, Peru, U.S.*	*Treatment settings*	*Behavior modification*

A major problem with assessing the impact of a therapy program has been the lack of a method which can measure and define the environment with which the patient is confronted. Jay Jackson has developed a questionnaire known as "The Characteristics of the Treatment Environment" (CTE), which fits the requirements of an instrument "sensitive enough to pick up subtle intrahospital differences between wards, and yet general enough to make quanitified interhospital comparison." This study employed the CTE questionnaire in a cross-cultural analysis designed to examine potential insights into mental hosptial practices in the United States. Two Peruvian mental hosptials were compared to fifteen mental hospitals in the midwest and western United States.

The CTE is based on five subscale aspects of treatment: 1) active treatment, 2) socioeconomic activity, 3) patient self-management, 4) behavior modification, and 5) instrumental activity. The questionnaire

determined the way in which the envrionment impinges on patients, rather
than on attempting to measure the potential therapeutic value of the
setting. "The emphasis in measurement is on the degree and direction of
constraints and freedoms imposed upon the patient and on certain element
of staff activity in this regard."

The two mental hospitals located in Lima, Peru were chosen as being
representative of the Peruvian treatment environment. The Larco Herrera
hospital was the largest mental hospital in Peru, and contained 1,200
patients, while the more recently established El Asesor hospital contained about 500 patients. Staffs at both hosptials answered the CTE in
September, 1966 and again in October, 1967. Final data were gathered
from 72 staff members at Laco Herrera, and 17 from El Asesor.

Results indicated that although the Peruvian hospitals did differ
slightly among one or two factors, the differences were not significant.
However, when the Peruvian hospitals were compared to the U.S. hospitals, statistically significant differences were discovered among four
of the five subscale categories. The one common factor was the degree
to which the staff attempts to influence, demand, or elicit certain
types of behavior (behavior modification). In general, the greatest
difference seemed to stem from the paternalistic relationship charater-
istic of Peruvian culture and treatment techniques. "The general atmosphere of paternalism manifests itself in some closer relationship with
patients, more restricted activities, and a directorship that assumes
the authority knows what is best for the patient." Thus, the Peruvian
mental patient tends to be more protected, and have less opportunity to
make choices than does his American counterpart. Regardless of whether
one treatment environment is superior over another, given the differences in the cultures each patient will be facing upon release, the
authors believed that "the successful cross-cultural use of the CTE has
increased its applicability to measurement of treatment environments in
mental hospitals.

BECKER, F.D., R. SOMMER, J. BEE, AND B. OXLEY   22

"College Classroom Ecology," *Sociometry,* Vol. 36, No. 4, 1973, pp. 514-
525.

KEY WORDS:

Man	Setting	Relational Concepts	Outcomes
College students	*Classrooms*	*Environmental arrangement*	*Social behavior*

The authors of this study suggest that there is a measurable relationship between a classroom's configuration and the behavior of its
inhabitants. Their purpose within the study is to qualify those effects
and sharpen the hypothesis by presenting specific ecological relationships between environmental arrangement and behavior.

Three separate experiments were performed in this study, the first
involving the observation of 51 classes ranging in size from 6 to 50 in
order to determine the effects on student participation and class size.
The second test involved the observation of 13 laboratory class settings

(both scientific and artistic) in order to establish a relationship between presentation technique and social interaction. A final experiment
was the administration of 282 questionnaires to three undergraduate
classes to determine the relationships of seating arrangement and position to academic performance, motivation, and degree of perceived similarity to the instructor.

The results of the first experiment failed to indicate a relationship between class size and student participation. Although the amount
of time increased in participation from large to small (2.5 versus 5.8
minutes) this increase was not significant when compared to overall
class time. The second experiment demonstrated a marked increase in
class participation and interaction when compared to the traditional
class arrangement. Forty-five percent of the student-teacher interaction times were more than 30 seconds and over 65 percent of the class
members participated in such discussions. The third experiment showed a
high correlation between seating position in a classroom and subsequent
class performance. Grades were related directly to proximity to the
instructor and significant relationships were found between the attitudes of the front and rear of the room. Seventy-five persons of the
members who occupied front seats expressed positive attitudes concerning
the class, for example, as opposed to 52 percent for those sitting in
the rear.

BECKER, F.D.   23

"Children's Play in Multifamily Housing," *Environment and Behavior,* Vol.
8, No. 4, December, 1976, pp. 545-574.

KEY WORDS:

Man	Setting	Relational Concepts	Outcomes
Apartment dwellers	Housing	Architectural arrangement	Children's behavior

This study is an analysis of children's play behavior in urban
environments and attempts to relate specific architectural features to
children and adult satisfaction with their recreational opportunities.
The author measured the user-satisfaction of small children, teenagers,
parents, residents and managers of seven New York housing developments.

The seven developments consisted of four low-rise apartment
complexes with planned recreational facilities surrounding the site and
three high-rise complexes with few planned recreational facilities.
Four data gathering instruments were used and consisted of 1) resident
interviews (n = 357), 2) questionnaire checklists (n = 591), 3) observations (n = 100), and 4) an interview of all apartment managers. The
interviews and questionnaires were designed to rate the residents'
overall satisfaction with the children's play areas and highlight those
facilities which were most satisfactory in each apartment complex. The
structured observations were utilized in order to give the author an
inventory of playground usage and catalogue the patterns of children
behavior in each type of play arrangement.

Some of the more salient findings of the study were that high-rise residents were less likely to be satisfied with the play areas (26%) than those living in low-rise apartments (43%). It was also found that the most popular forms of play equipment were moveable items such as swings and seesaws (46%), while the more static and "architectural" items such as climbing and adventure facilities, while accounting for 39% of the total equipment, were used only 5% of the time. It was shown that low-rise developments usually provided a wide variety of play situations and were able to accommodate a broad range of children groups. High-rise complexes, however, were characterized by a much more restricted range of play areas and very little space was devoted to active facilities. Most children in the high-rise situations were observed to be performing sedentary activities and using parking lots and stairwells for recreational outlets. A major deficiency of the high-rise complex was an inability on the parents' parts to supervise their children (38 - 50%) in comparison to a relatively high degree of supervision in low-rise apartment settings (54 - 89%).

BELCHER, J.C. AND P.B. VAZQUEZ-CALCERRADA   24

"A Cross-Cultural Approach to the Social Functions of Housing." *Journal of Marriage and the Family*, Vol. 34, No. 4, November, 1972, pp. 750-761.

KEY WORDS:

Man	Setting	Relational Concepts	Outcomes
Families	Georgia, Puerto Rico, & Dominican Republic	Housing attributes	Differential housing functions and preferences

The thesis of this paper is that family functions are influenced by the characteristics of the home, and that the dwelling is the arena of most interaction by the family as a group. An in-depth discussion of the various functions of housing in a cross-cultural context preceded description of a study conducted by the authors. The major hypothesis was that the physical structure and functions of the housing unit mediate in the functioning of the family as a group, and that a changing residential situation may serve to provide new functions for the family, and, ultimately, new bonds of stability.

Concepts relative to the discussion of the social functions of housing included a fixed-feature space, territorial needs, privacy, tradition as a predicator in the family social structure and subsequent house form; house form and socioeconomic status; the community-activity versus home-activity orientation dichotomy; and the social functions of housing when singularly viewed as "shelter" in a cross-cultural context. The authors maintained that there is a considerable amount of fragmentary information in the literature which suggests that most housing functions vary by socio-economic status as well as cross-culturally.

A systematic analysis was made, assuming that the functions people want their homes to fulfill were reflected in descriptions of their ultimate dream home. Comparisons of preferred dream homes of rural residents in Georgia, Puerto Rico, and the Dominican Republic were elicited in order to determine differential functions of housing. Climatic variations as variables that influence housing and its related social functions were not observed.

The data for this study were elicited through surveys in three different countries. The interview schedule was essentially the same on the housing questions in all three stages of fieldwork. Fieldwork was conducted in South Georgia in 1965, Puerto Rico in 1966, and the Dominican Republic in 1967. The subjects described in detail various attributes of their preferred dream home, including the construction materials, number of rooms in the house, overall size of the home, names or uses of the rooms, sanitary facilities, number of porches, the monetary value of their dream home, as well as other desirable characteristics.

The data concerning housing aspirations were augmented and thus made more complete through the examination of the manner in which selected family universal functions such as food preparation, eating, functions of the living room, and bathing were satisfied in the dwelling unit. All data were reported in percentage distributions across series of sub-sets of housing indices for each of the three countries.

The results indicated that many determinants of house form were intimately related to the variables addressed. There were culture-specific differences, but in general, the higher the level of living as a' group, the more functions were satisfied within the home. The manipulation of fixed-feature space in the housing forms for these three cultures was viewed as a mechanism for giving the family new functions as well as bonds that hold it together as a viable social unit.

The authors concluded that the simple provision of shelter can no longer be considered the major function of the dwelling unit, and that the configuration of the dwelling as a distinct architectural entity needs to be considered a possible molder of the structure and functioning of the nuclear family as the basic unit of society.

BETH-HELACHMY, S. AND R.L. THAYER, JR.   25

"Play Behavior and Space Utilization in an Elementary School Play Yard," *Man-Environment Systems,* Vol. 8, No. 4, July, 1978, pp. 191-201.

KEY WORDS:

Man	Setting	Relational Concepts	Outcomes
Children	*Outdoor play area*	*Space utilization; behavior patterns*	*Preferences*

In this study the assumption was made that not enough research exists that can aid in determining the extent to which children's behavior in school play yards is related to the physical environment, and conversely how much is related to other variables, such as attitudes of school authorities. Prior studies were discussed that focused on children's preferences and usage of playground facilities. Both obtrusive and unobtrusive measures have been utilized to gather data. The general consensus is that activities and preferences of play-space users were strongly related to and affected by the structure and form of the physical environment.

Objectives of the present study were to determine: 1) how children used the school play yard during recess time; 2) specific patterns of uses and activities among different ages and sexes; 3) which types of environmental conditions affect children's play behavior; and 4) the use of systematic observation techniques in a naturalistic setting.

A total of 65 primary and intermediate level pupils were observed during recess periods over a two-month period. Observations were recorded on a standard form designed to accommodate the collection of data during 30-second time intervals. Data pertaining to these seven behavioral categories of play were recorded: 1) location; 2) height; 3) enclosure (type of space play occurred within); 4) specific group activity; 5) social context; 6) body position; and 7) the number of children involved. A team of trained observers recorded various play behaviors in a manner that minimized disruption within the setting. Space usage was analyzed by measuring the percentge of time a child used a specific space type.

The study found: 1) no significant age or sex differences were detected in use of the various space types; 2) children spent the largest amount of time (39%) playing on hard surfaces; 3) children spent 43% of their time in structured activity; 4) only 6% of children's time was spent playing with the permanent play equipment; and 5) turf areas were not used as much as initially expected. Additionally, children's individual behavior "orbits" were monitored. These orbits were: 1) sex differentiated; 2) the physical setting; and 3) the child's grade level. It was concluded that in future efforts designers should strive to develop the outdoor spaces near classrooms to achieve maximum utility. Also, facilities intended for use by children of various ages should be located in neutral zones with equal accessibility to all children.

BICKMAN, L., A. TEGER, T. CRKRIELE, C. MCLAUGHLIN,   26
M. BERGER, AND E. SUN

"Dormitory Density and Helping Behavior," *Environment and Behavior,* Vol. 5, No. 4, December, 1973, pp. 465-490.

KEY WORDS:

Man	Setting	Relational Concepts	Outcomes
Students	Dormitory	Density	Helping behavior

Two separate studies indicated that dormitory density has a measurable effect on the helping behavior of students. The authors suggested that the more dense a dormitory (the most people per structure) the more likely are the students to be distrustful, unfriendly, uncooperative, and less helpful.

Three degress of dormitory density were defined for structures at both the University of Massachusetts and the University of Pennsylvania. At the former, two female dorms averaging 528 students were defined as high density, 8 housing 166 each were classified as medium density, and 37 housing 58 each were classified as low density. In each type of dorm, preaddressed, stamped envelopes were randomly placed in public areas at specific times in order to measure the effects of helping behavior (tendency of residents to mail the letter with no reward). In addition, a random sample of 126 residents of all three dorm types was administered an attitude questionnaire that tapped a measure of their living habits, self-perception, perception of their neighbors, and perception of their dorm environment. A similarly designed experiment using housing arrangments at the latter university was performed that involved a mixed sex sample and an additional overt-helping test.

In the first study it was found that 58% of the letters were returned in the high density dorm, 79% in the medium density, and 88% in the low density, which indicated a high degree of correlation between density and helping behavior(X^2 = 17.48, p< .001, df = 2). It was also shown through the attitude questionnaire that residents of the low density structures were more trustful (88% versus 48%), more likely to help a same-sex confederate in need (53% versus 23%), and more likely to report vandalism (82% versus 33%) than residents in a high density situation. In the second study at Pennsylvania the basic results of the first experiment were replicated with a 63.6% return rate for letters in the high density dorms compared to 77.4% and 90.6% for medium and low density environments (X^2 = 5.10, df = 2, p< .001). A semantic differential test of attitudes in this latter study also indicated significant variances in degrees of social interaction between the three populations. The authors concluded that the results of the two studies supported the hypothesis that helping behavior and positive social attitudes are related to lower dormitory densities. They found that residents in low density environments tended to be more gregarious, friendly and trustful, and rated their living area as more open and comfortable.

BELL, B.D.   27

"The Impact of Housing Relocation on the Elderly: An Alternative Methodological Approach," *International Journal of Aging and Human Development,* Vol. 7, No. 1, 1976, pp. 27-37.

KEY WORDS:

Man	Setting	Relational Concepts	Outcomes
Elderly	*Urban area*	*Housing type*	*Satisfaction*

This article examines the impact of housing relocation on the patterns of interaction and life satisfaction of a sample of older, married adults living in an urban area of central Arkansas. Interviews were conducted with 115 individuals residing in congregate housing and with a matched sample of 105 persons living in independent residential units. The congregate housing settings were chosen principally on the basis of two characteristics: 1) the congregate situation was not of an institutionalized nature and 2) the majority of residents (70 percent or more) were sixty years of age or older. Included among the congregate housing were retirement "hotels," high-rise complexes, and communal residences (i.e., private housing given over to occupancy by five or more "families").

In this study the data were examined in regard to four hypotheses. First, congregate living situations will yield significantly greater frequences of interaction than will be observed in the case of independent residences. Second, significant differences will be observed in the type of interaction characterizing the congregate and independent residential arrangements. Third, that there will be a direct relationship between residential duration and both interaction and life satisfaction. The fourth and final hypothesis was that there will be a direct relationship between residential duration and both interaction and life satisfaction.

In regard to the first hypothesis, there appears to be no appreciable variations in interactions regardless of housing context. On the basis of the data, hypothesis one was discounted.

In the case of hypothesis two, significant differences were expected by the author in the type of interaction characterizing the congregate and independent settings. No significant differences, however, were found to exist in terms of character of interaction displayed in the five areas investigated (interaction, status, employment status, and life satisfaction). That is, the respondents appeared to interact in the same areas with approximately the same degree of regularity, regardless of residential context.

Hypothesis three, on the other hand, treated more directly the effect of extended residence in the congregate setting. A direct relationship was suggested between residential duration and both interaction and life satisfaction. In the case of interaction, the correlation between residential stay and interaction was inverse rather than direct. That is, interaction frequencies tended to decline over time for congregate dwellers. The correlation between residential duration and life satisfaction moved in a negative direction. Here, too, the trend was toward a gradual decline in satisfaction over time. Thus, there would appear to be little in the way of empirical support for hypothesis three.

Finally, significantly higher life satisfaction for the congregate as opposed to the independent dwellers was found. The data generated from the study, however, indicated significantly higher estimates of life satisfaction for independent residents. This relationship held true regardless of age, sex, status, health, and the residential duration of the respondent.

The study conducted by Mr. Bell clearly suggested that the congregate housing setting had insufficient resources to offset the objective as well as subjective decrements of old age.

BISHOP, D.   28

"User Response to a Foot Street," _Town Planning Review_, Vol. 46, No. 1, January, 1975, pp. 31-46.

KEY WORDS:

Man	Setting	Relational Concepts	Outcomes
Pedestrians	England/street	Environmental setting	Public response

Four months after the opening of a model "foot street" in the
Portsmouth, England, Commercial Road Shopping Centre, a survey of pedestrians was made by students from a nearby polytechnic institution using
a personal interview technique. Its purpose was to obtain information
from the general population in their perceptions of the shopping centre
in terms of environmental and retail attractiveness as well as general
information concerning the interviewee.

Eligible persons were 12 years of age and over with 5,173 interviews made on four different survey days. Weekend visitors tended to be
younger, in larger groups, own more cars and shop more than weekday
pedestrians. In both groups, over 90% were in favor of the pedestrian
district or "foot street." Safety, litter, crowding, convenience and
appearance problems were all perceived by the public as improved;
although the noise, litter, and crowding improved only slightly. Accessibility deteriorated for both cars and buses. Suggestions by the
public for further improvement included added parking, extension of the
foot walk, added toilets, and covered areas.

The main conclusion of the study was that the majority of public
users were in favor of the pedestrian precinct. It was also concluded
that using a real, full-scale model which could be altered, as opposed
to a permanent change, allowed the public to have a say concerning their
environment without extravagant spending on the part of government.

BLAUT, J.M. AND D. STEA   29

"Studies of Geographic Learning," *Annals of the Association of American
Geographers,* Vol. 61, No. 2, June, 1971, pp. 387-393.

KEY WORDS:

Man	Setting	Relational Concepts	Outcomes
Children	Cross-cultural	Geographic learning; cognitive mapping	Map reading photo interpre- tion

Cross-cultural and developmental research on the untaught mapping
abilities of children aged five through ten suggests that mapping behavior is a normal and important process in human development, and that
map learning begins long before the child encounters formal geography
and cartography. The major objective of this study was that new strategies might be developed and incorporated in early geographic education
and learning. Certain findings on geographic learning of young children
were reported from a study of the ontogeny of environmental behavior.
It was found that mapping behavior was an essential part of human development and an important activity of preschool children. A simple mapping communication model was conceptualized as possessing the following
functions: a) information regarding distance, direction, and physical
features (cognitive map); b) the discrimination of a subset of mapping
processes; and c) criteria developed to determine what constitutes a
cognitive map (scale, projection, and abstract signs). The translation
of these aspects are largely iconic, or pictorial.

Black and white vertical aerial photographs were used as the principal instrument for testing the map-reading abilities of preliterate
children. A set of environmental experiments were designed to elicit
the untaught mapping abilities of school-entering children in three different cultures.

In the first group of tests, 107 beginning first graders in
Worchester County, Massachusetts, and 20 children in Rio Piedras, Puerto
Rico were asked to interpret aerial photographs (of their respective
areas) by naming and pointing to features which they recognized.

The median age of the children was six years, five months. All but
2 of the 107 children perceived the photograph to be a downward view of
the landscape and identified at least 2 unlike features on it. The mean
score of correct identification was 6.4; incorrect identification were
few; and Puerto Rican scores were comparable to Worchester scores. It
was concluded that the photo-interpreting ability of first graders in
the test populations was confirmed.

A second study with 19 first graders from the same area determined
children's ability to handle two other mapping problems: a) to draw
meaningful map signs and understand the meaning of these signs; and b)
to perform a simulated map-using task which showed children's ability to
deal with a set of map signs as a single system. Sixteen of the group
performed the entire series of tasks. The remaining children performed
all but the final task. This provided further evidence that first-
graders can read iconic maps and some evidence regarding their ability
to deal with rather abstract maps.

A similar set of tasks were performed by a group of 58 first grade
children in the isolated mountain community of St. Vincent, West Indies.
For this group, the mean score (based on a photograph of their region)
was 1.4 correct identifications. A pair of Gestalt questions were also
asked of this group. The results paralleled the first two studies and
supported the initial cross-cultural model. The results of the overall
research indicated that the basic components of mapping behavior are
displayed by young children in three different cultures. Results also
suggested that geographic learning is clearly, if not completely, dis-
tinguishable from social learning.

BOALT, C.   30

"Living Habits in Residential Areas," *Build International,* Vol. 3, No.
9, September, 1970, pp. 257-260.

KEY WORDS:

Man	Setting	Relational Concepts	Outcomes
Households	*Sweden*	*Household design features*	*Preferences/satisfaction*

In order to better design homes that are in accordance with the
wishes of its inhabitants, it is important to determine exactly how
people use their living spaces. Through the Stockholm Municipal Housing

Companies, the National Swedish Institute for Building Research conducted a study of living habits in five southern suburbs of the city. The intent of the study was to "contribute to the development of swifter, cheaper, and surer methods of studying users and their dwellings and also to throw some light on a number of questions of topical interest."

The survey was conducted in two stages. Stage one contained such factors as the stock of available flats, the users, the area (to a limited extent), and the users' evaluations of a number of features of the dwellings and the area. A total of 2996 questionnaires were sent to all households in the area, and replies were obtained from 2893. Evaluations were made on a seven-point scale for a number of separate features of each residence, as well as for the residential area in general. Data were also gathered on the composition of the household, the age, sex, and occupation of each of its members, spatial features and standard household equipment, and the ownership of cars, washing machines, and television sets.

Stage two was more concerned with noise disturbance and the accessibility of services, and utilized personal interviews of 568 housewives. A number of questions were also raised regarding the use of the dwelling, the tendency to move, and requirements with regard to a building.

Results indicated a wide range of preferences and characteristics. First, the length of time a household had occupied their home did not seem to have any effect of their ratings. Thus, it would not appear that the wish to move is a suitable measure of the way in which residents feel about their dwellings. Secondly, kitchens which received the highest ratings were characterized by their large total area, equipment arranged along two adjacent sides, and a dining area near a window. In terms of sanitary facilities, long, narrow bathrooms were not well-received. The desire was for daylight and a separate water closet. Living room and storage facility ratings increased with size. Of all the features of the dwelling, however, sound insulation was assessed the most unfavorably. Approximately two-thirds of the households mentioned some noise outside the dwelling as disturbing. Finally, it was found that people are most inclined to invest in holidays and recreation, followed by the dwelling, furniture and other fittings, food, clothes, and domestic appliances.

BOOTH, A. AND J.N. EDWARDS   31

"Crowding and Family Relations," *American Sociological Review,* Vol. 41, No. 2, April, 1976, pp. 308-321.

KEY WORDS:

Man	Setting	Relational Concepts	Outcomes
Residents	*Congested households*	*Density and crowding*	*Personality/family relations*

Many studies regarding the crowding of animals have indicated that dense living conditions may result in a disorganization of mating practices, as well as a disintegration of other types of social behavior

such as maternal care and infant development. The potential significance of these findings formed the impetus for this Toronto-based study
which examined the effects of household and neighborhood crowding on the
relation between spouses, parents and children, and between children.

The initial population pool for the study consisted of those census
tracts in Toronto with the "potential in yielding a large number of
families residing in dwellings in which the number of people exceeded
the number of rooms." Of the 17,000 households screened with the
desired demographic characteristics, 522 wives and 344 husbands were
actually interviewed. Thus, the end result was basically a sample of
blue-collar families residing in very compressed as well as relatively
open home and neighborhood conditions.

Multiple regression analyses of the gathered survey data were used
in order to screen out the effects of age, socioeconomic status, education, and ethnicity from crowding consequences. Comparison of those
families who lived in crowded settings with those who resided in
uncongested home and neighborhood environments allowed the authors to
examine a wide range of family interactions.

An examination of the results revealed that crowding is virtually
unrelated to family relations. Of the 84 possible relationships considered, only 11 were statistically significant, with one of those
suggesting that crowding enhances family ties. In general, household
congestion was found to have only a small influence on the frequency of
sibling quarrels, little effect on how often parents struck their
children, and no adverse consequences on marital relationships. In
fact, a negative relationship appeared to exist between people per room
and the number of arguments between spouses. The authors, therefore,
concluded that crowded conditions produce modest and infrequent con-
seqences, which indicated that "we have a good deal of flexibility in
housing people in dense environments."

BORN, T.J.   32

"Variables Associated with the Winter Camping Location of Elderly
Recreational Vehicle Owners in Southwestern Arizona," *Journal of Gerontology,* Vol. 31, No. 3, May, 1976, pp. 346-351.

KEY WORDS:

Man	Setting	Relational Concepts	Outcomes
Elderly	Campgrounds in Southwest U.S.	Choice of site; differ- ences between groups	Camping preferences and behaviors

During the winter of 1973-74, 580 elderly recreational vehicle
users were interviewed in six private trailer parks and four public
camping areas in Southwestern Arizona and Southern California. The ten
sites represented four types of camping environments: 1) highly developed, relatively expensive, urban private trailer parks; 2) moderately
developed, relatively inexpensive, rural private trailer parks; 3)
authorized, developed, free federal campgrounds; and 4) unauthorized,

undeveloped, free public camping areas. The author cited prior research that set precedents with regard to the characteristics, behavior, and preferences of these retired persons and their use of recreational land in the southwest during the winter season. The primary objective of this study was to test for significant differences in socioeconomic backgrounds among four subgroups within the total sample in order to explain the use of certain campgrounds over others.

A combination interview-questionnaire was used to gather data. The questionnaire addressed a wide array of issues pertinent to the exploration of subjects' motivations, preferences, needs, and desire relative to their choice of campgrounds. A random sampling technique was used in the private trailer parks coupled with "geographical sweeps" in order to obtain the most diverse, representative, and yet balanced sample from each subgroup. Many questions were directly related to characteristics and behaviors of the female.

The analysis initially studied relationship between 21 variables. Subsequent analytical steps identified six variables of primary importance: 1) per person annual income; 2) woman's age; 3) woman's total adult preretirement camping experience; 4) value of mobile quarters and tow-vehicle; 5) man's education; and 6) percentage of year lived in the recreational vehicle. Analysis of variance was utilized to identify respondent characteristics most strongly associated with the type of site selected for winter residences.

The results indicated that variables associated with subjects' financial situations were very important in explaining winter camping behavior. Several financial and certain noneconomic indicators of females were also strongly associated. It appeared that the woman's role was critical in camping decision making. A large percentage of subjects from all subgroups cited a need for certain physical amenities such as water and sewage facilities. More that 50 percent of the total sample of public land campers had no permanent, conventional home since many had no economic choice. The author found that the subjects were persons who subsisted on low incomes, sought to maintain independence, possessed camping orientation that often stretched back for many decades, and persons who looked to public land resources to help them in their post-retirement years.

It was concluded that recreational vehicle camping by the retired elderly in the desert Southwest posed challenges for the public land manager. Incomes among these campers were relatively low and were lower at undeveloped campgrounds. Fees and amenities provided must accommodate the behaviors, preferences and economic backgrounds of these users.

BRAND, F.N. AND R.T. SMITH   33

"Life Adjustment and Relocation of the Elderly," *Journal of Gerontology,* Vol. 29, No. 3, May, 1974, pp. 336-340.

KEY WORDS:

Man	Setting	Relational Concepts	Outcomes
Elderly	*Housing*	*Relocation*	*Decreased social networks*

The problem of life adjustment as related to relocation of the
elderly was reported in this study. Social and psychological data on
the study population were obtained by interview and health status data
by physical examination. A group of 68 subjects aged 65 years and over
who experienced forced relocation because of urban renewal showed higher
scores of maladjustment (as measured by a Life Satisfaction Index) than
a control group of 69 nonrelocated subjects.

The authors reviewed studies which focused on the adverse effects
of environmental change on the elderly. Some of these reports have
shown a high mortality rate among the elderly due to forced relocation.
Few prior studies have analyzed forced relocation based on factors
related to satisfaction levels and coping behaviors regarding altered
social networks of these persons. The study and control groups were of
similar age, sex, race, education, rent payments category, and health
status. The Chi-square measure of association with Yates correction was
used to test for significance.

Relocation had more adverse effect on females than males, and black
subjects adjusted better than white subjects. The study showed that for
subjects in poor health, life satisfaction was lower among the relocated
group. Social interaction of the elderly with the built environment was
a critical factor in successful life adjustment following relocation.
When compared to the nonrelocated group, relocated subjects were
generally less active and had fewer social contacts.

These findings suggested that involuntary changes did disrupt carefully evolved social networks and foster potentially harmful effects on
the elderly. The physical and social stress caused by forced relocation
was viewed as a potential hinderance to the elderly, even in cases where
persons are moved to newly built environments.

BRONZAFT, A.L. AND D.P. MCCARTHY   34

"The Effect of Elevated Train Noise on Reading Ability," *Environment
and Behavior,* Vol. 7, No. 4, 1975, pp. 517-525.

KEY WORDS:

Man	Setting	Relational Concepts	Outcomes
Children	*Elementary school*	*Noise; read- ing skills*	*Decrease*

Relationships were studied between the noise of passing elevated
trains in a large metropolitan city and reading skills of children in a
nearby elementary school. When reading skills of children whose
classrooms were parallel to the train tracks were contrasted with scores
of children whose classrooms were on the quieter side of the building,
students on the noisy side scored significantly lower. Reading scores,
obtained from school records of the 1971, 1972, 1973 and 1974 academic
years of classes on the east side (noisy side) of the building were compared with matched classes on the west side (quiet side) of the school.
A 2 x 2 x 2 analysis of variance performed on the word knowledge and
reading comprehension scores of the children in the matched classes
yielded a significant effect for location at the .01 level.

The child's tendency to block out both relevant and irrelevant cues when taught in a noisy classroom and the loss of actual classroom teaching time as trains passed by the setting were seen as two possible explanations for depressed reading scores. Noise measures taken in the classrooms while in session (without train noise) were at 59db. When a train passed, the noise level rose to 89db. It was determined that classes were disrupted on the noisy side of the building every 4 1/2 minutes for an interval of 30 seconds by the noise of the passing train. Since the present findings supported earlier findings on the harmful after effects of excessive exposure to noise, urban agencies and designers were urged by the author to seek new methods for reducing potentially harmful urban noise.

BROOK, R.M. AND P. KNAPP   35

"Effects of Residential Evaluation and Rehabilitation Placement on Children's State - Trait Anxiety," *Journal of Clinical Psychology,* Vol, 32, No. 1, January, 1976, pp. 57-59.

KEY WORDS:

Man	Setting	Relational Concepts	Outcomes
Children	*State residential facility*	*State-trait anxiety/relocation*	*Adjustment levels*

This study focused on the environmental stress experienced by children who were removed from a home and placed in a Kansas state institutional setting. Children entered either a 90-day evaluation unit or a more time-extended residential rehabilitation program. Because physical isolation coupled with confrontation with a new living environment may be dangerous or ego-threatening to a particular child, it was hypothesized that admission to the program would cause a heightened level of anxiety and that this level would remain stable during the adjustment period as the child became more knowledgeable about the new environmental setting. Literature was reviewed which suggested that levels of trait anxiety remain stable across time and are conceptualized as transitory emotional responses.

Data were elicited from 20 males and 4 females. Six of the subjects were runaways or had severe reading handicaps. The mean age was 13 years, 10 months and mean IQ was 93.5. The 10 evaluation-unit and 14 rehabilitation program children averaged a 5.7 grade level. A series of tests were administered to each child under standard conditions. Initial data were elicited upon the child's arrival at the youth center, and a follow-up test occurred 21 to 25 days after arrival. Additonal data were obtained through the regularly scheduled evaluation and assessment procedures.

The hypothesis that trait anxiety would remain relatively stable across time ($t = .64$, $p < .50$) and the prediction that environmental isolation and new social controls initially would be experienced as ego-threatening ($t = 3.24$, $p < .01$) were strongly supported. Thus, the confrontation with the new physical environment and social setting did not have a significant impact on the children studied according to this

set of variables and measures. Results of a two-tailed test for correlated measures that compared admission and post-test scores provided
support for the use of these particular measuring tests to explore
anxiety tendencies in children. The author concluded that future studies should explore item analysis with regard to subjects' sensitivity
to particular levels of state-trait anxiety and the type of threatening
environmental and social situation.

BROOKES, M.J. AND A. KAPLAN   36

"The Office Environment: Space Planning and Affective Behavior," _Human
Factors,_ Vol. 14, No. 5, October, 1973, pp. 373-391.

KEY WORDS:

Man	Setting	Relational Concepts	Outcomes
Office workers	_Office land-scaping_	_Spatial arrangements_	_Perception_

It has been suggested that although open office arrangements
improve interpersonal communication, a serious lack of privacy and an
increased degree of perceived noise will offset any social gain attained
by these types of work environments. The authors of this article have
proposed that current modes of designing office landscapes are insuf-
ficent in dealing with complex attitudinal behavior that is the direct
result of these environments.

The study focused on a large U.S. retail firm that transferred
three departmental functions occupying 30,000 square feet of conventional, cubicle-type office space to an open-plan office setting that
relied only on plantings, differences in finishes, and moveable furnishings for visual and physical separation. One hundred twenty workers
from the three departments were administered questionnaires immediately
prior to and nine months after the office transfer. Semantic scales
were used to elicit the respondents' perceptions of both work environments and their perceptions of what would constitute an "ideal" environment. Attitudes were measured in terms of such indices as noise,
privacy, color, shape of spaces, and sociability.

Results of the test indicated that the subjects perceived a significant loss of privacy and an increase in noise perception in the open
office environment. Although group cohesiveness and sociability indices
increased after the transfer, no direct link between the environmental
setting and this improvement could be found. The authors found that
there was no evidence to suggest that the subjects used the new design
to adapt their surroundings to their work, but rather that supervisory
policy of the departmental heads was the controlling variable of
workspace layout. They concluded that the advantages gained in the
group interactions after the use of the open office concept were counterbalanced by the deficiences of added visual and noise distractions.

"Outdoor Recreation as a Function of the Urban Housing Environment,"
Environment and Behavior, Vol. 6, No. 3, September, 1974, pp. 295-345.

KEY WORDS:

Man	Setting	Relational Concepts	Outcomes
Residents	_Urban neigh-borhoods_	_Outdoor recreation_	_Attitudes/perceptions_

A study was conducted in three inner-city neighborhoods in Baltimore to determine how residents used the outdoor spaces. Two neighborhoods were low-income. Data were gathered of potential use both to residents and city agency personnel in making decisions for the development of specific sites within the study areas. The authors provided a detailed physical and social profile of the ten sites studied within the three overall neighborhoods. The paper presented detailed results from four of the study sites.

A number of data gathering techniques were employed. These methods were designed to reveal objective data as well as the attitudes and perceptions of residents which might influence behavior. Methods included: 1) a drive-around census of outdoor activities throughout the study area; 2) walk-around census of outdoor activities at selected sites within the area; 3) outdoor activity diaries kept by a sample of residents at each of these sites; and 4) interviews and perceptual studies with residents. For analysis purposes activities were classified as in transit, housework, recreation, around the house, parks and playgrounds, and other. A series of classifications were also developed for each of the other three data gathering techniques.

In these various sites and through a series of unobtrusive observations it was revealed that the streetfront was a major locus of recreation. Many more persons engaging in recreation activities used sidewalks and steps than those persons using both backyards and playgrounds. Residents tended to spend leisure time on the street front even when backyards or adjacent playgrounds provided them with clear alternatives. Playgrounds were used well below their technical capacity. Fewer people of all age groups were observed participating in recreational activities in playgrounds than immediately around the dwelling. The third neighborhood analysis was middle-income. This area differed from the two low-income neighborhoods in several respects. The majority of outdoor sitting behavior and children playing tended to be away from the street front. Backyards were well utilized for recreational purposes and interior block parks were extremely popular.

The discussion of findings focused on home-based recreation, convenience, safety, facility-based recreation, and a series of planning implications and objectives for the neighborhoods that were studied. The authors concluded with a brief history of planning efforts in these areas and possible explanations for the ineffectiveness of prior planning and urban design strategies.

"The Effects of Voluntary and Involuntary Residential Mobility on Females and Males," *Journal of Marriage and the Family,* Vol. 35, No. 2, May, 1973, pp. 219-227.

KEY WORDS:

Man	Setting	Relational Concepts	Outcomes
Adults	*Urban areas*	*Residential mobility*	*Mental disorder; preferences, intentions*

    The authors explored the effects of voluntary and involuntary staying and moving upon family members with special attention given to differential effects upon both sexes. Their conceptualization of residential mobility linked decisions involving past and current preferences and intentions with subsequent mobility behavior. Also, choice of moving was seen as an important link in the moving process. Two possible inhibiting conditions that might arise when a family is unable to carry out its preferences were the family becomes either an involuntary stayer or an involuntary mover.

    Data were utilized from a two-way national survey of residential mobility pattern and preferences in U.S. metropolitan areas. Interviews with approximately 1500 households in 43 SMSA locations were conducted in the fall of 1966, using a multistage probability sample. During the fall of 1969, 1,561 individual interviews were again carried out with members of the original households. The measure of involuntarism relied upon the relationship in residential mobility reported by the same persons during the period between 1966 and 1969. A systematic analysis interrelating spatial mobility choice in 1966 and in the period between these two years resulted in four categories: 1) voluntary stayers (chose to stay), 2) involuntary stayers (chose to move but stayed), 3) involuntary movers (chose to stay but moved), and 4) voluntary movers (chose to move). Fifty three percent of the sample were voluntary stayers. For the other three categories the figures were 18.4, 14.4, and 14.2 percents, respectively.

    A series of detailed hypotheses were tested which were based on these four residential mobility-choice categories. The major variables and the results of the analyses were presented in a series of tables which addressed interrelationships between subjects-households' voluntary and involuntary residential mobility behavior and 1) organizational participation, 2) the frequency patterns of visiting with families in and out of the neighborhood, 3) the perception of neighbors regarding various demographic-behavioral indicators and 4) various social psychological measures, such as alienation, unhappiness, mental disorders, and poor physical health.

    The results suggested that for both males and females moving, whether involuntary and/or voluntary, decreasing formal organizational participation had little effect upon 1) informal social relations either within or outside the neighborhood of residence, 2) how respondents perceived their neighbors, and 3) alienation, unhappiness, suspected mental disturbances, and poor physical health. However, recent residential mobility experiences had a marked effect upon the mental health of

females. The authors concluded that further research should consider
other measures of adaptation to new environments, including frequency of
moves and more detailed measures that focus on dimensions of the immediate residential environment.

BYROM, C.   39

"Privacy and Courtyard Housing," _The Architect's Journal,_ Vol. 151, No.
2, January, 1970, pp. 101- 106.

KEY WORDS:

Man	Setting	Relational Concepts	Outcomes
Residents	_Low cost housing, England_	_Privacy Social inter- action, Activity space_	_Preferences satisfaction_

    The desire for housing which provides a great deal of privacy is on
the upswing. Citing a lack of information regarding "the kind and
degree of privacy tenants require, the positions in which it is most
wanted, and to what extent attitudes toward privacy are influenced by
such factors as age, sex, stages in the life cycle, and social class,"
the author investigated what privacy meant to a particular low cost
housing group in Dundee, England.

    The study took place in a courtyard housing development covering
four acres and with a density of forty-eight persons per acre. Most of
the units were designed for families with two or three children. The
average head of household was 50 years old for men and 48 for women.
Only 4 percent of the inhabitants were less than ten years old.
Information was gathered based on a single interview with each house-
holder and on observations made by an architect during the interview.
Final results were obtained from 45 households regarding two particular
aspects of privacy: 1) freedom from disturbance by noise and 2) freedom
from overlooking or being overlooked by others.

    Results indicated that many doubts regarding the suitability of
courtyard housing to provide adequate privacy were unfounded. While
emphasizing that privacy is valued, subjects responded that there were
very few intrusions on the privacy of their individual houses. The
enclosed outdoor courtyard space was well liked as offering an activity
space free from interference by those living nearby. When compared with
similar results from two related studies, the author concluded that
"privacy, defined in broad terms as freedom to live one's own life
without marked interference or intrusion by those living close by, is an
important element in many people's lives. Consequently, it cannot be
ignored by designers and others concerned with the field of housing."
Further emphasizing the current gaps in information regarding how, why,
or where privacy is important, the author stated that the evidence from
this study confirms the importance of privacy, particularly in connec-
tion with outdoor activity spaces.

"Interpretability of Graphic Symbols as a Function of Context and
Experience Factors," *Journal of Applied Psychology,* Vol. 60, No. 3,
June, 1975, pp. 376-380.

KEY WORDS:

Man	Setting	Relational Concepts	Outcomes
Engineering students	Machinery usage	Communication	Symbol recognition and meaning

Ten graphic symbols designed by Henry Dreyfuss Associates for Deere
and Company farm and industrial machinery were tested for ease of
interpretation in context and in isolation. The international "language
free" modes of communication that are needed in the human factors interface with machines was seen as not being expressed by design features
alone, not even by supposedly "self-evident" design unless augmented
with meaningful graphic symbols, which provided these machines with a
visual voice. The author hypothesized that the symbols, as a set, would
be more often correctly indentified in context and by those subjects
with relevant prior experience with the symbols.

Thirty male college seniors comprised the subject group. The subjects differed in their level of familiarity with the machinery. All
subjects completed a questionnaire and then were divided into two equal
groups defined as either context or no-context. Subjects in both groups
were instructed to identify the meaning of each out-of-context symbol.
Context group subjects were presented with a drawing of the cab interior
of a piece of heavy equipment typical of the kind in which these symbols
would be used. The numbers that appeared on instruments and controls in
the drawing were used to have each subject locate each symbol in the cab
and to identify the meaning of the symbols in context. The 10 symbols,
accompanied only by their respective numbers, were presented as 35-mm
slides. Seventeen of the 30 subjects were rated as experienced with
these specific types of machines and 8 of them were in the no-context
group. Of the 13 inexperienced subjects, 7 were in the no-context
group.

Results indicated that the 10 graphic symbols maintained the same
relative order of difficulty under both context and non-context conditions, a difficulty which ranged from 100% correct responses to only a
few correct responses. These findings supported the authors view that,
in general, graphic symbols were most meaningful within an appropriate
context, or frame of reference. Although this constraint was recognized
by Dreyfuss, it has all too frequently been overlooked by the more en-
thusiastic proponents of symbology.

Within the context of this study, differences were evident in favor
of a greater ease of recognition of symbols by subjects having specific
operational familiarity with heavy industrial equipment. In designing
nonmachine systems, the author believed that a more explicit and precise specification of user background was required than merely the
assumption that the operator will have reasonable intelligence and an
average amount of mechanical knowledge, coupled with some degree of
familiarity with the operations of agricultural and industrial equip-

ment. Empirical validation of effectiveness was viewed here as a necessary step in the symbol development process.

CAMERON, P., D. ROBERTSON, AND J. ZAKS   41

"Sound Pollution, Noise Pollution, and Health: Community Parameters,"
Journal of Applied Psychology, Vol. 56, No. 1, February, 1972, pp. 67-75.

KEY WORDS:

Man	Setting	Relational Concepts	Outcomes
Families	*Urban areas*	*Sound pollution; noise, and health*	*Noise judgements*

In order to explore urban parameters of noise and sound pollution, a random sample of 2,130 Detroit and 496 Los Angeles families were interviewed. The major questions for which preliminary answers were sought were these: 1) Does exposure to considerable amounts of sound affect health? 2) What are the normative frequencies of exposure to loud sounds in our society among adults? 3) Where are such sounds encountered? 4) How loud are such sounds, and how frequently are they encountered? 5) Is the home or work situation affectively more pleasant from a sound standpoint, and are there sex differences in these judgements?

The cultural context, the relative nature of sound, the environmental context, and causes of noise related problems and decreases in work efficiency were reviewed by the authors. A questionnaire was developed that included a battery of coded questions pertaining to characteristics of sound and noise. Both the noise and sound questions were coded via the same method. The code home, work, or other applied where sound or noise was experienced. The source of sound was coded for machine(s), children, traffic, planes, and music. Loudness was coded one for the jet plane response, two for in between the jet plane and the propeller plane, three for the propeller plane, and so on down to eight for less than five or six people talking together. Frequency was coded as every day, 5 times per week, and less than 5 times per week. Responses were recorded and analyzed by a trained group (interrater reliabilities averaged 94% agreement) using the same diagnostic categoies employed by the Public Health Service in their annual survey of the nations' health status.

An adult member of the family reported the health status and sound-noise exposure of all family members. Reports of 2,906 females and 2,854 males were obtained. Two types of data was elicited, "self-report" (personal health assessments) and "other reports" (assessments of others in the household). A chi-square test was used to measure the comparability of both kinds of data although they were not treated as independent of one another. Of the 1,144 comparison, 17 were statistically significant at the .01 level. A series of tables presents both self and other report data.

Results suggested that: 1) there may be an association between sound exposure and an increased prevalence of both acute and chronic illness; 2) males are exposed to noise more frequently than females; 3) machine, plane and traffic sounds are usually considered noise, while musical and childrens' sounds are usually considered sound; 4) the sounds encountered at work are usually considered noise while sounds experienced at home are usually considered sound; and 5) about a quarter of the women and a third of the men in our society are annoyed by sounds on a regular basis.

CANTON, D. AND R. THONE   42

"Attitudes to Housing: A Cross Cultural Comparison," *Environment and Behavior,* March, 1972, pp. 3.

KEY WORDS:

Man	Setting	Relational Concepts	Outcomes
Students	Laboratory	Culture	Preferences

Since the inception of the "International Style" in architecture, there has been a steady erosion of cultural differences between buildings. The current assumption is that we adapt to the Western paradigm on which most buldings are designed and the perception and understanding of the forms of buildings will be similar across cultures. Attitudes to the built-environment were expected to be highly sensitive to cultural differences and thus provide a good test of the assumptions underlying a cultural design.

First year architecture students at the University of Strathclyde, Glasgow, Scotland, and the University of Sydney, Sydney, Austrilia, were shown a series of slides and told to evaluate each on a scale from 0 to 10 with the higher score indicating more satisfaction. In addition, each subject was requested to complete three open-ended sentences for two of the illustrations. Responses fell into four categories: 1) disapproval; 2) approval; 3) description; and 4) no answer.

The specific expectation of this study was that people living in a particular environment of dwelling types might be more sympathetic to those to which they are accustomed. Generally, the students had a preference for old terrace housing, and a particular type of non-traditional individual house. In comparison, the Glasgow students preferred the traditional, individual bungalow style of house typical in Australia. Neither showed any preference for the average modern row houses in Britain.

CARP, F.   43

"Short-Term and Long-Term Prediction of Adjustment to a New Environment," *Journal of Gerontology,* Vol. 29, No. 4, July, 1974, pp. 444-453.

KEY WORDS:

Man	Setting	Relational Concepts	Outcomes
Elderly	Residences	Relocation	Person-congruence

For the elderly, a change in residence can have deleterious effects, including a rise in death rate, or it can have generally favorable results via increase in adaptive capacity. These results are always determined by the interaction of personal and situational factors. An analysis of group trends following relocation yielded valuable insight as to the nature of the environment-person interaction which determines types of change experienced by the elderly. An analysis of within-group differences showed that this adjustment was not uniform among residents - particularly when the new environment was held constant. Differences in both the immediate period following the move and over an 8 year period were measured to account for "sleeper" effects of the change. It was hypothized that behaviors which occur during a "honeymoon" phase of community development are not those that occur when a situation stabilizes and the community is firmly established.

The appropriateness of the concept of person-situation congruence was tested in predicting the adjustment of elderly persons who formerly occupied substandard housing to a new, nine story complex designed for 200 persons. Terms in a prediction equation varied according to the following criterion: 1) residents' happiness; 2) acceptance by peers; and 3) adjustment to the living situation as rated by the staff. Each of these measures are maintained as a separate index. Residents' capacity for independent living was a requisite for admission to the new complex. Predictor items were identified for each applicant prior to any data collection concerning relocation. The independent living predictor of adjustment focused on a series of situational variables: 1) income; 2) marital status; 3) chronological age; 4) education; 5) job level; 6) self-concept; 7) physical disability; 8) confusion-memory loss; 9) senility; and 10) neurotic-type complaints.

Activity and involvement predictors were: 1) desire for organized activity; 2) desire for informed activities; 3) attendance at religious services; 4) social disengagement; 5) cognitive age identification; and 6) feelings about the future. Sociability predictors measured: 1) active social preferences; 2) number of close friends; 3) number of friends in persons' youth; 4) relationships with parents in youth; 5) current attitude toward others; 6) levels of friendliness to others; 7) undue interests in others - noisiness; 8) nervousness; 9) fastidiousness - good taste; and 10) overall likability. An assessment of each applicant's overall Person-Situation Congruence was made at this point.

For the study of the impact of the new facility on residents, data were collected for 352 legally qualified applicants. At the time of final data collection, 133 members of the original in-moving group remained in the faciliy. Interviews were conducted with each resident at predetermined intervals. Stepwise multiple regression analysis was used to establish a prediction equation between the total group of 27 predictors and data collected for in-movers at the end of the first year in the new building for each of 3 adjustment criteria. The study was cross-validated over time.

Results indicated that the equation developed for each criterion with short-run data also accounted for an appreciable amount of the variance in scores at the end of the 8-year interval. It was found that while the basis for person-situation congruence must be specified according to the definition of adjustment - the attributes of short-run adjustment are also conducive to long-term adjustment if the same definitions apply to both phases. Further study by the author will elaborate on issues dealing with the long-term validity of this system

which accounted well for three measures of short-run adjustment to a new
environment.

CARP, F.   44

"Ego Defense or Cognitive Consistency Effects on Environmental
Evaluations," *Journal of Gerontology,* Vol. 30, No. 6, November, 1975,
pp. 707-711.

KEY WORDS:

Man	Setting	Relational Concepts	Outcomes
Elderly	*Housing*	*Hypothetical-actual reloca-tion; defense; cognitive con-sistency*	*Increased disso-nance associated with moving*

This study focused on some of the major discrepancies that were
felt to exist between elderly residents' and observers' evaluations of
specific residential environments. These differences in evaluation were
hypothesized as being the resultant effort of ego defense or dissonance
reduction on behalf of subjects. Also, where these psychodynamic processes were evident, old persons' evaluations were hypothesized as
becoming more negative when their efforts to move to a more desirable
setting met with success. Both empirical findings and the assumptions
held by many researchers lend credence to the fact that elderly persons
in substandard or otherwise inadequate living environments tend to evaluate their situations more positively than do data collectors or other
observers. A number of studies have documented the environmental settings of elderly residents and no matter how favorable the subjects
rated their own situation, the authors of these prior studies usually
managed to conclude that the area was either "deteriorating" or in a
"run down" condition. This study sought to determine the nature and
effects of these discrepancies in perception.

A low rate of negative evaluations of "an obviously poor (living)
situation may reflect strong ego threat or cognitive dissonance." It
was assumed that this dissonance existed because "the individual's behavior (was) inconsistant with his self-concept." One way for a person to
reduce such feelings of inadequacy is to deny that negative aspects of
their living environment exist.

The hypothesis was tested with the original group of applicants
after they relocated to a different project. In both cases data were
collected on nearly all applicants declared eligible by the local public
housing authority. The subject groups possessed similar demographic
characteristics. Each elderly person was interviewed twice for each of
the two major time periods. This took place before persons moved
(before prospective tenants saw the new building). Evaluation of present housing was included in both interviews. Data were analyzed via
the distinction between persons who were offered new apartments and
those who were not.

In both cases (the effect of being an applicant and the effects of being offered an apartment) the hypotheses were supported. Applicants who were selected to move to new housing became more negative in their evaluations of housing than applicants who were not selected to move. It was found that the subjects who knew they might be provided the opportunity to move into a better living environment were capable of a more realistic assessment of their present living situations. For persons who were not offered an apartment, evaluations of housing were only mildly negative when compared to the other group. No one from this group rated his or her housing as "good", "very good", "poor" or "very poor". The shift in evaluations for persons at time 2 were significant (X^2 = 81.59, p< .001 for Group 1 and X^2= 33.39, p< .001 for Group 2).

It was concluded that when these individuals were locked into a demeaning living situation which was inconsistent with self-concepts, one way employed to reduce the dissonance was to change perceptions of the present living environments. These persons were highly responsive to a real change, even though the change was "totally hypothetical" - before any actual move took place.

The findings suggested that verbal information can underestimate the negative aspects of a situation where the person is unable to relocate, and that it is unjustifiable to assume that an elderly person is in a benign situation merely because they are verbally assured by the person that his or her living situation is "all right".

"Impact of Improved Living Environment on Health and Life Expectancy," *The Gerontologist*, Volume 17, Number 3, June, 1977, pp. 242-49.

KEY WORDS:

Man	Setting	Relational Concepts	Outcomes
Elderly	Housing	Housing conditions	Health

It is assumed that improving the quality of a living environment will also improve the health and longevity of the elderly. This article is a study of that hypothesis and an attempt to relate the effects of environmental influences to social and genetic factors.

The study surveyed approximately 550 tenants and applicants to a new senior citizen complex and 150 comparison respondents in the general population of San Antonio, Texas. The interviews were performed before a move to the complex, after a one-year interval, and at the end of eight years in order to demonstrate the continuing health benefits of improved housing conditions. The respondents were asked to describe their own attitudes concerning their health as well as their ability to function within the complex. In addition, specific medical problems were assessed in terms of frequency, length, and seriousness of illnesses among the complex residents and the comparison group. Death rates were also compared between the groups and detailed logs of activities were maintained.

The author concluded that modern and improved living environments are a positive health factor to the elderly, both in the short and long term. Major factors in the improved health status of people living in the new complex were better temperature control, safer stair and flooring design, cleaner surfaces, and better security control. The author also suggested that the new environment was also more supportive in terms of social contact with friends and peers and an improved self-esteem in the ability to live a more independent life style. The overall indices of health and mortality were approximately 10 to 20 percent higher for residents living in the new complex than for those of the comparison group.

CARR, S.J. AND J.M. DABBS, JR.   46

"The Effects of Lighting, Distance and Intimacy of Topic on Verbal and Visual Behavior," _Sociometry_, Vol. 37, No. 4, December, 1974, pp., 592-600.

KEY WORDS:

Man	Setting	Relational Concepts	Outcomes
Female students	_Laboratory_	_Distance/ lighting_	_Behavior_

This study measured the relationships between behavior and indices of intimacy. These indices were defined for this test as distance between interviewer and subject, lighting level of the environment, and topic of discussion. The authors proposed to reinforce existing hypotheses that state a direct correlation between intimacy indices and behavior.

Forty female students were interviewed separately by one of two female graduate students. A baseline interview was conducted in a neutrally lit room, with the two women sitting opposite a desk at a distance of 3 feet. The subject of the baseline interview was nonpersonal and general. The degree of direct eye contact was recorded during the experiment as was the length of subject response to each question. Each subject was then introduced to the second interviewer and moved to a second room in which a much more specific and personal interview was performed. The 40 settings of the second experiment were divided equally between the four combinations of bright/dim and close (1.5 feet versus 8 feet) variables. Again, recordings were kept of eye contact and length of responses. After the second interview, a questionnaire was administered concerning each subjects' perceptions of lighting, intimacy of topic and proximity to interviewer.

The final questionnaire established that subjects were able to discriminate the differences in lighting level and distance from baseline to experimental interviews. A high correlation was found between lighting levels and visual behavior in that the subjects found a dimly lit room inappropriate for the conduct of the intimate interview. No such correlation was found with respect to the distance variable which was thought to be the result of having same-sex pairings of interviewer and subject. In general, the authors concluded that verbal and visual behavior is dependent upon specific environmental factors (such as lighting) in relation to the context and content of a specific social encounter.

CARTER, D.J. and B. WHITEHEAD   47

"A Study of Pedestrian Movement in a Multi-Story Office Building,"
Building and Environment, Vol. 11, No. 4, 1976, pp. 239-247.

KEY WORDS:

Man	Setting	Relational Concepts	Outcomes
Office workers	*Office building/ England*	*Circulation communication*	*Cost-effectiveness*

In contrast to the studies done regarding the circulation in hospitals, schools and police buildings, the literature regarding cir-culation in large office buildings is limited. The basic elements involved in any system of office planning revolve around the space requirements for given activities, and the communication links between units. Yet, a major problem with collecting data on the circulation of an office is that if it is gathered from one particular environment it is not necessarily applicable to other office building problems. Thus, this study carefully chose a setting that was of modern design, built for a purpose, contained a wide range of activity types, and employed people with duties and contacts among a number of interrelated depart-ments. The aim of the research was to arrive at suggestions for mini-mizing movement costs within the site studies, and, given the research constraints, to generalize these findings to the extent possible.

The Liverpool Divisional Headquarters of British Railways London Midland Region provided the setting for the study. The ten-story building contained approximately 700 staff members, working in a mixture of open plan and large and small cellular offices. A survey was taken over a period of two weeks during June, 1975, designed to determine the association between activities and to study the vertical circulation system. In addition, two observers randomly visited 20 predetermined observation points covering all activities during the working day, and recorded inter-activity journeys. The observed data was then reduced to a more manageable matrix of activity patterns by eliminating activities generating little movement and by combining adjacent activities which performed similar functions.

In terms of circulation within this particular building, it was discovered that activities within four of the six major departments were all located on the same floor. Thus, it may be concluded that the importance of communication links between activities had been appre-ciated during the building design stage. Small changes in layout and organization were suggested to reduce circulation costs even further.

In general, it was determined that the costs of movement of people in a building will ultimately be greater than all of the initial, main-tenance, and other running costs of the building taken together. Thus, although the best design solution must also be able to deal with a multitude of design considerations and restrictions, a building which matches the operational characteristics of its organization is of major importance.

CHEYNE, J.A. AND M.G. EFRAN   48

"The Effect of Spatial and Interpersonal Variables on the Invasion of
Group Controlled Territories," *Sociometry,* Vol. 35, No. 3, September,
1972, pp. 477-489.

KEY WORDS:

Man	Setting	Relational Concepts	Outcomes
Students/ general public	*Shopping mall and university*	*Territorial intrusion*	*Use of space*

A neglected but important aspect of human territorial behavior is
the appropriation of space by two or more individuals. Two separate
studies investigated a number of variables which influence an individual
who intrudes upon group-shared space. Reaction to intrusion, which can
occur frequently in an urban environment, is not the subject of these
studies. Instead, the stress experienced by the intruder was recorded.

Both studies hypothesized that male-female and male-male pairs
would defend territory more effectively than would female-female pairs.
The first study involved students from the University of Waterloo in
Ontario, Canada. The experiment involved 60 students having to walk
around two of their fellow students in a narrow corridor leading to a
reception deak. Many varibles were tested including dyads of male-male,
female-female and male-female combinations and three corridor situations: dyads conversing, dyads facing away from each other, and use of
2 wastebaskets (as a control) in place of the dyad.

The 60 subjects were randomly selected by observers who watched for
students walking alone who approached the dyads in order to reach the
reception desk. Results indicated that the female-female dyads who were
not conversing deterred the smallest number of subjects from walking
between them. In fact, the proportion of subjects who walked between
the two non-conversing females was greater than the number who walked
between the two inanimate wastebaskets. It was also found that the
ratio of male to female subjects was 5:1 which may have influenced the
outcome of the study. A male is apt to be reluctant to disturb any
male-female pairing and would probably not come between two conversing
females. However, the male subjects would welcome a chance to walk between two idle females.

In the second study, the same type of corridor situation was
created by using a second level balcony-style walkway, 137 inches wides,
overlooking the lower mall in a large Toronto shopping center. Placement of the student dyads in this walkway left 63 inches of space to
pass by and 41 inches of space between the conversants. All pedestrians
walking by who weren't walking in groups were recorded by several observers.

When statistically analyzed, the results indicated that all student
dyad combinations were able to deter pedestrians from walking between
them when they stood 40-46 inches apart. However, no combination of
male or female dyad could control the number of territorial intrusions
when the distance between the conversants was increased to 52 inches,
although male-female dyads fared better than the others.

In hallways ranging in width from 86 inches to 137 inches, the cue indicating to others the nature of interaction between two people seems to be the absolute distance between the conversants. However, those same distances between conversants did not keep others from interposing themselves when the corridor or walkway was in a place of public business.

CIMBALO, R.S. AND P. MOUSAN   49

"Crowding and Satisfaction in a Banking Environment," *Psychological Reports,* Vol. 37, 1975, pp. 201-202.

KEY WORDS:

Man	Setting	Relational Concepts	Outcomes
Banking customers	Bank	Crowding	Satisfaction

This study sought to determine reliable situational and behavioral correlates for customers in a banking environment. It was hypothesized that looking around at other patrons, rather than looking at one's own material would be a predictor of dissatisfaction.

Eighty-six randomly selected customers (53 males and 33 females) participated in the study. Four types of looking responses were recorded: 1) looks around, 2) looks at customers, 3) looks at own material, 4) looks at teller. In addition, information on the total amount of time spent in bank, number of people in the bank at the time, number of comparisons, a measure of each participants satisfaction level (measured on a 9 point scale), and frequency of visitation were ascertained by observation and direct questioning. A Pearson correlational analysis was performed on these variables.

A small but significant Pearson correlation was found between satisfaction and number of customers in the bank ($r = .22$, $p < 0.01$). Satisfaction was not significantly correlated to any of the "looking" measurements. There was a significant but low correlation between satisfaction and number of customers in the bank for males but not for females. In addition subjects spent more time and looked around more frequently at other customers the more crowded the bank became ($r = 0.61$ and $r = 0.58$, $p < 0.001$). This data was seen as supporting Summer's findings and providing support for a personal space explanation of satisfaction in banking environments. In other words, violation of personal space was seen as a causal variable of dissatisfaction in this environment.

"Location of Static Gatherings in Pedestrian Areas: An Exploratory
Study," *Man-Environment Systems,* Vol. 7, No. 1, January, 1977, pp.
41-54.

KEY WORDS:

Man	Setting	Relational Concepts	Outcomes
Pedestrians	*Urban plaza*	*Proxemics/ behavior settings*	*Space usage*

     It was hypothesized that individuals and groups often appear to
make nonrandom choices with respect to spaces they occupy or deliberately avoid. Studies were reviewed which focused on the ways people
distribute themselves in outdoor, large-scale urban settings. The
author outlined five general conclusions based on the summarization of
prior empirical and intuitive research approaches to this subject. A
number of factors were cited that were viewed by the author as inherent
in successful urban public spaces. These factors included the elements:
visibility, comfort, protection, identity, and amenities. The present
study utilized the location-point concept that approaches group spatial
behaviors as sets of cartesian coordinates or in terms of distance to a
reference point. The study setting was described as possessing four
basic factors. Distances and orientations of persons were analyzed
according to these categories. Observations reported were conducted in
a pedestrian plaza situated in one of the shopping centers in Canberra,
Australia.

     A series of 500 aerial photographs were taken of the space showing
all persons simultanteously in a variety of behaviors and locations.
The population using the plaza was dividied into three groups: sitting,
standing or walking individuals. In the final sample 352 cases of static gatherings were coded and analyzed. A number of variables were
viewed relative to the observed behaviors.

     It was found that the locations chosen by spontaneously formed con-
versational gatherings were systematically related to: 1) physical
objects within the environment; 2) the configuration of pedestrian traf-
fice lines; and 3) to distribution of static persons in the plaza.
Locations selected by standing groups also were significantly affected
by gathering size and sex composition. A number of possible explanations for the results were presented. It was concluded that the placement of a gathering of persons in relation to various environmental
amenities such as trees, walks, benches and lamps indeed influence behavior. Designed objects served as delineating, or "space-establishing"
elements that create architectural spaces which determined the interac-
tional domain of persons in urban public spaces.

"The Experimental Control of Littering," *The Journal of Environmental Education,* Vol. 4, No. 2, Winter, 1972, pp. 22-28.

KEY WORDS:

Man	Setting	Relational Concepts	Outcomes
Human	*Theater/ Campground*	*Reinforced incentives*	*Behavioral adjustment*

The authors of this article argued that reinforcement incentives can be a powerful behavioral tool in modifying a population's performance in the environment. Specifically, they presented the hypothesis that incentives given to people in a litter control program will result in much higher ratings of effectiveness than any other reinforcement system.

Two tests were performed and documented in this study to prove the above thesis. The first was a 14-week experiment in a children's theater in which a variety of anti-littering devices were compared to an initial baseline period. The devices ranged from the installation of extra trash cans and signs to the payment of each patron for the return of any waste generated during the move. The payments were either 10 cents or a free movie ticket. A second experiment was performed over a two-week period in a national park by placing 160 pieces of trash in a camping area, and, given a no-incentive and incentive anti-littering program, measure the cleanliness of the area over time. The incentive program in the second week offered children a modest gift for their participation.

The authors found that the addition of even a small incentive in the two anti-litter programs dramatically improved the cleanliness of the two environments. It was found that the introduction of the dime reward in the theater raised the return rate of litter to 94 percent compared to 57 percent for anti-littering instructions and 16 percent for the baseline. A similar effect was found in the campground experiment, leading the researchers to conclude that the most appropriate methods of litter control were those which incorporated some tangible piece of reward for subsequent clean behavior.

CLARK, R.J. 52

"Environmental Technology - A Modern Myth," *Build International,* Vol. 7, No. 3, May-June, 1974, pp. 265-277.

KEY WORDS:

Man	Setting	Relational Concepts	Outcomes
Families	*Households/ England*	*Noise*	*Avoidance reaction aural preference, environmental design*

The author contended that "it is seldom appreciated that problems which involved the total environment inevitably involve a multiplicity of disciplines, and, unless there is some focal point which can be seen by a manufacturer or by a builder, then the total solution will always be obscure and only when a problem becomes dramatic will any positive steps be taken towards the required result." For instance, the functioning of a central heating system is dependent upon a series of separate manufactured items, designed to come together as a system, but each with its own set of rules based on different criteria. The result is that such factors as the house in which we live - a fundamental component in determining a person's quality of life - is the last factor to be considered by a technologist designing one component of a central heating system. Clark further believes that "the fact that money could be saved by anticipating a problem through planned inquiry is seldom recognized."

In order to illustrate the importance research can play in reducing the discrepancy between a technologist's design and consumer preference, the author cited a recent study done involving the problem of noise eminating from central heating system pumps. The investigation determined the effects of noise and vibration of the pump and the acceptable limits of noise from such units. Approximately 400 households were visited in order to include a cross-section of the British population in terms of age, sex, environments, and background.

Results showed that a large majority of people were altering their environment by moving from one room to another. Measuring the efects of the noise on behavior proved to be difficult, since they appeared to be closely related to psychology, economics, and similar factors. In at least 70% of the cases, however, there was no question of people becoming accustomed to the noise when it predominated in the frequency range 500-1000Hz.

On the basis of these findings, the complete design of the pump eventually revolved around the noise factor. Clark believed that this example illustrates the importance of research working in conjunction with technology. "The most important factor in the end analysis was that by virtue of this work new and better materials were adopted, new and better techniques of system design were developed, and a fuller understanding of the needs of this particular industry were achieved."

COHEN, S., D.C. GLASS AND J.E. SINGER   53

"Apartment Noise, Auditory Discrimination, and Reading Ability in Children," *Journal of Experimental Social Psychology,* Vol. 9, September, 1973, pp. 407-422.

KEY WORDS:

Man	Setting	Relational Concepts	Outcomes
Children	*Apartments*	*Noise*	*Reading ability*

The relationship between a child's auditory and verbal skills and the noisiness of the child's home environment were studied. Expressway traffic along a heavily traveled highway in New York City was the principal source of noise. Initial decibel measurements taken at predetermined locations around and within a high rise housing development that bordered the highway permitted use of floor level as an index of noise intensity in the apartments.

Several prior studies have documented delayed effects of stressors other than noise. The major purpose of this study was to report systematic data in support of an hypothesized association between environmental noise stress and subsequent behavior, focusing on the child's length of residence in the housing development as influencing certain behavioral outcomes. The two lines of research influencing this design were: 1) a child reared in a noisy environment eventually becomes inattentive to acoustical sounds, which leads to impaired auditory discrimination; and 2) the delayed effects of noise on impaired reading performance in children needed to be analyzed in a real setting as opposed to laboratory observations.

Elementary school children who resided in four 32-story buildings of an apartment complex were tested via a battery of standardized measuring devices. Noise decibels were measured as decreasing from lower to upper floors of each building. The major independent variable was the length of time a child has resided in the complex. It was predicted that the longer a child was exposed to uncontrollable noise, the more he/she learned to "filter" both relevant and irrelavant sounds out of awareness. The child's reading ability suffered in this process.

The final sample consisted of 54 children (second, third, fourth, and fifth graders). Tests administered for reading, including sight vocabulary comprehension, and reading total were analyzed as percentiles. Auditory discrimination was measured via the Wepman ADT. Additional tests were given for exploratory purposes. An audiometric test coupled with pertinent social background data elicited from parents were incorporated as control variables. Testing was conducted in classroom settings. The sample was divided into two criterion groups according to those children who had lived in the complex 4 years or more, and children who had resided there 3 years or less. Data analysis procedures included correlations between variables for each criterion group and series of regressions to determine amounts of variance attributed to specific testing measures.

Children living on the lower floors of the 32-story buildings showed greater impairment of auditory discrimination and reading achievement than children living in upper-floor apartments. Auditory discrimination appeared to mediate an association between noise and reading deficits, and length of residence in the building affected the magnitude of the correlation between noise and auditory discrimination. Additional analyses ruled out explanations of the auditory discrimination effects in terms of social class variables and physiological damage. Results were interpreted as documenting the existence of long-term behavioral after effects in spite of noise adaptation. The authors concluded that this demonstration of post-noise consequences in a real residential environment supplemented laboratory research showing the stressful impact of noise on behavior.

"Increasing Litter Depositing Through the Use of Positive Conditioned Reinforcement," *Journal of Environmental Systems,* Vol. 6, No. 2, 1976-77, pp. 173-180

KEY WORDS:

Man	Setting	Relational Concepts	Outcomes
Students	*Cafeteria*	*Reinforcement signage*	*Behavioral change*

Conditioned reinforcement is the manipulation of a subject's response by a predetermined cue or prop. In this study, the researchers hypothesized that, by the installation of a relatively inexpensive cue system, litter control in a cafeteria could be significantly increased.

A university cafeteria was the experimental setting for this study. This area served approximately 1500 persons between the hours of noon and 1 p.m. each day and utilized a self-serve and clean-up system of food service. The researchers were interested in the number of persons who failed to clean their litter after eating with no signage cues and then again with various environmental props. Records of the number of uncleaned trays were recorded without cues for 28 days and then a series of reinforcement systems were instituted for the next 22 days. Interobserver agreement was approximately 90 percent during the experiment and variances within the data were statistically controlled.

The results confirmed the original thesis of the authors in that cleanliness of the cafeteria area increased after the institution of a reinforcement system. During the first 28 days of no control, approximately 68 percent of the users failed to clean their litter. After the installation of a sign above the trash area and the manual distribution of thank you cards, the clean-up rate rose to over 90 percent. A series of variations in the signage system was tried during the experiment, each of which showed an increase in trash control after a reinforcement cue was installed. The authors concluded that the low cost of the cues ($7.00) was very cost effective both in economic as well as environmental terms. Even though the tokens were in themselves useless, a marked increase in positive behavior was achieved through their usage.

"Ramps or Stairs: The Choice Using Physiological and Biomechanic Criteria," *Applied Ergonomics,* Vol. 3, No. 4, December, 1972, pp. 195-201.

KEY WORDS:

Man	Setting	Relational Concepts	Outcomes
Human	*Stairs and Ramps*	*Physiology*	*Conservation of energy*

It has been proposed by ergonomic standards that the trade-off point between the efficiency of ramp usage and stair usage is 20 degrees (below 20 degrees, ramps are physiologically more efficient). The authors of this article proposed that by studying the biomechanic effect of stair and ramp usage, this standard could be either reinforced or refuted.

Eight males between the ages of 20 and 28 were used as subjects for the tests. They varied in height and weight and were physiologically healthy. An experimental set of adjustable ramps and stairs was constructed and each subject was connected to equipment which measured his oxygen intake, heart rate, and knee and ankle movements. The subjects were told to scale the equipment at their own paces and records of each were recorded to measure the effect of differing lengths of scaling times. Ramp and stair angles were adjusted between 10 and 30 degrees, with riser height varying between 4 and 6 inches. The Shinno equation of leg extension was used as a composite measure of the subject's knee and ankle flex.

Given the fact that all subjects were in good health, the results of the test showed that the physiological cost of climbing stairs is always less than a ramp of equal slope. It was found, using Shinno's formula, that biomechanically the ramp of equal inclination is easier to negotiate, but with a greater expenditure of energy. The 6 inch riser was found to be, at least of healthy subjects, the most desirable height for both energy efficiency as well as speed. The standard of the 20 degree trade-off slope appeared to the authors to be only valid in relation to limb angle.

CORSON, J.H. AND C.W. CRANNELL   56

"Simple Reaction Time to Cutaneous Temperature Stimuli," _Canadian Journal of Psychology,_ Vol. 2, No. 5, October, 1970, pp. 305-310.

KEY WORDS:

Man	Setting	Relational Concepts	Outcomes
Students	_Laboratory_	_Tactile thermal response_	_Warm-cold response variation_

Studies in the past which have involved the measurement of simple reaction time to temperature stimuli applied directly to the skin have not been practical due to the fact that when a probe is used, the subject is first aware of the touch, and must "hold back" his response until he is aware of the temperature. This study, however, utilized an apparatus which overcame this problem and measured time differences for various skin temperature stimuli.

The subjects consisted of six male and six female undergraduate students who took part in six one hour laboratory experiments. The mean and median response times were recorded for 2.8, 5.6, and 8.3 °C each above and below the measured skin temperature, with only one temperature being used per session. A subject sat behind a plywood barrier and placed his arm through a hole fitted with a cloth sleeve so that any vision to the apparatus on the other side was prevented. The wrist was then placed lightly on a set of tubes which moved, and gave the impression that there was a uniformly textured metal surface moving transversely beneath the wrist. After the subjects were given white noise through a set of foam-padded earphones to reduce the likelihood that sound cues were being picked up, several temperature trials were given and their corresponding response times recorded.

When the mean and median response times were compared across the range of temperatures, no differences between male and female responses was noted. However, it was discovered that response time was clearly slower for stimuli which was above normal skin temperature than for stimuli which was below, with response time decreasing as the deviation from skin temperature increased. While discussing these findings in relation to contemporary theories of thermal sensitivity, the authors concluded that there can now be little doubt that a difference does exist between response times for warm and cold.

CRUMP, S.L., D.L. NUNES, AND E.K. CROSSMAN   57

"The Effects of Litter on Littering Behavior in a Forest Environment," *Environment and Behavior,* Vol. 9, No. 1, March, 1977, pp. 137-146.

KEY WORDS:

Man	Setting	Relational Concepts	Outcomes
Recreation-ists	*Park*	*Presence of litter*	*Littering behavior*

The purpose of this research was to determine the relationship between litter in a forest environment and the resulting littering behavior of people who use the environment. Specifically, the study attempted to measure the validity of Heberlein (1971) and Finnie's (1973) hypotheses that a clean area tends to influence behavior in a positive fashion and prevents littering.

Data were collected over four consecutive Fridays in a picnic area in Uinta National Forest, Utah. Two specific picnic sites were studied with an approximate user population of 42 on each day observations were made. On the first and third test days, all trash was removed from the picnic sites at 8 a.m. and an inventory of litter was conducted after 9 p.m. On the second and fourth Fridays, 60 pieces of litter were spread through both areas before the park was opened and an inventory was again conducted after it closed that day. An observer was stationed at the sites on the first and last test dates in order to count the user population and make observations concerning the behavior of people entering both experimental conditions.

Contrary to the hypotheses of Heberlein and Finnie, the results of this study indicated that a site littered with the 120 pieces of trash at the start of a day had less litter at the end than the sites which began with none at all. In the nonlittered areas, for example, a total of 93 pieces of trash during the first week and 149 during the third were inventoried at the close of the park. In the areas on the second and fourth test dates that were intentionally planted with 120 pieces of trash, only 55 and 53 pieces were counted after 9 p.m. Since the number of persons using the park did not vary from one test to the next and no distinct behavioral variables were noted, it was postulated by the authors that the presence of the trash was a major factor in the subsequently more clean behavior of picnikers. The studies of Heberlein and Finnie were carried out in urban environments, while this experiment was performed in a natural setting, which could explain the drastic differences in results of this and past tests.

CUTLER, S.J.   58

"The Availability of Personal Transportation, Residential Location, and Life Satisfaction Among the Aged," *Journal of Gerontology*, Vol. 27, No. 3, July, 1972, pp. 383-389.

KEY WORDS:

Man	Setting	Relational Concepts	Outcomes
Elderly	*Rural community*	*Transportation; residential location; life satisfaction levels*	*Lack of personal transportation, low life-satisfaction scores*

In the context of a community which had neither public nor commercial transportation facilities, the life satisfaction of the elderly was related to differentials in the availability of personal transportation modes. It has become increasingly clear that the elderly do not have equal access to the advantages provided through personal transportation nor are they equally able to surmount the obstacles that exist in its absence. The literature cited by the author focused on research conducted in urban areas concerning life-satisfaction factors, and he stressed the need to learn more about these critical variables in the context of a smaller, rural community.

Two hypotheses were tested: 1) Without public and commercial transportation, life satisfaction levels would be higher among elderly persons who have personal transportation as opposed to those who do not have access to personal transportation; and 2) A stronger relationship exists between the unavailability to personal transportation and life-satisfaction levels for persons whose residences are farther from destination points than among those person whose residences are more proximate.

The respondent sample consisted of 121 females and 49 males whose median age was 74. Two community features critical to the analysis were that dependence on personal transportation for mobility was high; and the community possessed a high degree of centralization since most facilities-amenities were located within a one-quarter mile radius of the center of Oberlin, Ohio. The primary independent variables were based on differentials in availability of personal transportation and differentials in residential location in terms of distance from the center of the city. A questionnaire sought an array of information concerning persons' residential location, mobility, availability of personal transportation, and life-satisfaction levels.

Results did not support hypothesis one: 58% of the aged with transportation available to them were rated high on the life satisfaction index, while only 37% of those without transportation had high scores (p< .02). The second hypthsis was supported: in that older persons living more than one-quarter of a mile from the center of the city, 73% of those with transporation and only 22% of those without transportation were rated high on the index (p< .01). Lower levels of life satisfaction characterized the elderly persons who did not have personal transportation available to them and who also lived further from service facilities and amenities of the community. The analysis confirmed the significance of the need to consider how the variability of environmental factors relates to the morale, mobility, and life-satisfaction adjustment of the elderly.

Transportation differentials were known to be most critical when environmental barriers could not be overcome by these persons and when no viable alternatives were available. Also, life-satisfaction scores for persons who were of low socio-economic status or in poorer health were lower than for others. The author concluded that further study would be necessary in order to empirically assess the significance of persons' frequency of participation in social activities, walking as a viable transportation mode, and as studied via a longitudinal research design that focuses on these issues relative to the elderly in small rural communities.

DAVIDSON, A.I.G., H.G. SMYLIE, A. MACDONALD, AND G. SMITH   59

"Ward Design in Relation to Postoperative Wound Infection: Part II," *British Medical Journal,* Vol. 1, January 9, 1971, pp. 72-75.

KEY WORDS:

Man	Setting	Relational Concepts	Outcomes
Patients	*Hospital wards*	*Treatment setting design*	*Cross-infection control*

In the first study the environment was designed in such a way that two beds were in one block and were divided only by an area used for scrubbing up. Ventilation consisted of a slow continuous air flow which turned over at a rate slightly less than two air changes per hour. The ward was one of a suite of four and was ventilated by a continuous air exchange system which achieved a rate of 10-20 air changes per hour. The requirements for participation in this investigation consisted of an operation involving an incision through clean, prepared skin which took place in the actual operational theater. Final data was gathered on 1000 patients.

In 145 cases (14.5%) the operations were complicated by some degree of wound infection. It was noted that the incidence of all types of infections was found to be greater after potentially dirty rather than after clean operations, regardless of the environment. Nevertheless, the infection rates were lower after the move to the new environment. Of primary significance was the decline of all major secondary infections which arise from staphylococcal cross-infection. Although the authors warn that the reduced incidence of wound infection cannot be totally explained by improvements in ward design and environmental control, they concluded that "from the information available regarding the bacteriological status of the wound environment in the operating theater it seems that the ward has been the site of major improvement. Differences in the pattern of secondary infection seem to confirm that the new ward environment had a beneficial effect in reducing the hazard of staphylococcal cross-infection."

DAVIES, A.D.M. AND M.G. DAVIES   60

"User Reaction to the Thermal Environments: The Attitudes of Teachers and Children to St. George's School, Wallasey," *Building Science,* Vol. 6, No. 2. July, 1971, pp. 69-75.

KEY WORDS:

Man	Setting	Relational Concepts	Outcomes
Teachers	*Schools/England*	*Thermal comfort*	*Environmental*
students		*Solar design*	*preference*

This study solicited opinions from teachers and students regarding the thermal environments of two different building structures within the same school complex. Attitudes pertaining to the conventionally heated main building were compared to those regarding the partly solar heated annex.

Separate surveys were devised for the staff and children. The children responded during the months of June and July, while the staff completed their surveys in July and February so that comments could be obtained under both summer and winter conditions. Two surveys were given the children. The first was open-ended in order to allow them to mention any feature of the environment which might be important, but which the experimenter might not have anticipated. The fixed-alternative second survey presented them with the opportunity to indicate which standard sets of answers came nearest to their opinions. The teacher's summer survey combined closed-ended questions using a five point rating scale with a number of open-ended questions. The winter survey was similar in nature, but contained additional questions designed to follow-up on the leads learned during the summer survey.

Results indicated that "as a teaching environment for the physically active child, the majority of staff preferred the Main School (p < 0.01). For personal comfort no significant difference of rating was found." However, it was generally felt among staff and children that the interests of solar heating had resulted in too many sacrifices. The corridors were dark and tunnel-like, and the solar wall blocked a desirable view. With the present design, the warm, stable temperature of the annex was generally recognized and appreciated during the winter months. In hot, sunny weather, however, the room either got too hot or the windows had to be opened, resulting in greater susceptibility to interruptions by noise. Thus, the authors concluded that in this case the problems of temperature control and ventilation have not been solved, but that the problems tended to arise from the operation of this particular building rather than from any inherent feature of solar design.

"The Effects of Music and Task Difficulty on Performance at a Visual Vigilance Task," *British Journal of Psychology,* Vol. 64, No. 3, August, 1973, pp. 383-390.

KEY WORDS:

Man	Setting	Relational Concepts	Outcomes
Students	*Laboratory*	*Detection latencies, noise, vigilance*	*Visual vigilance*

Relatively few studies have attempted to examine the effects of varied auditory stimulation upon vigilance performance. Those that have been undertaken have usually assessed the environmental variables in terms of one response measure. The authors believe that this tendency is potentially misleading since it is quite probable that different response measures reflect different processes, all of which are not influenced by the environmental variable in question. Consequently, the major aim of this study was to examine the manner with which music affects different indicators of vigilance performance.

Forty undergraduate students (20 men and 20 women) were tested individually and in isolation. Ten subjects were each randomly assigned to each of the four conditions of the experiment: Noise Easy (NE), Noise Difficult (ND), Music Easy (ME), and Music Difficult (MD). Their task was to respond to changes in the brightness of a light located 1.38 meters away. In the "easy" conditions, the brightness of the light changed in increments of 0.9 foot candles (fc), while in the "difficult" conditions the increments were 0.5 fc. At five second intervals the light flashed on for one second. Subjects were seated in front of a panel with two separate response keys and were instructed to press the "no" key whenever they thought the standard light had been pressed, and the "yes" button when a light of different brightness flashed. At the same time either musical selections or white noise was being played at an averge of 75 db.

Response measures were taken for correct detections, commission errors, detection latencies, and db values. Results indicated that when the task is easy, music has virtually no effect on the detection rate. However, when the task is difficult, music improves the detection rate substantially. In both music and noise conditions there is a tendency for performance at the difficult task to decline with time.

In terms of detection latencies, "under difficult conditions in noise situations, detection latencies significantly increase as a function of time, while in no other condition does time at work exert a significant influence on detection latencies." In addition, the mean number of errors was greater for music than for noise, although not to a significant degree. Finally, values of db are significantly lower when the task is difficult than when it is easy. Thus, the authors concluded that the best interpretation of the results is probably in terms of arousal. "It thus seems that either a more intense signal, or music, can cancel out the decrement in detection latencies."

"The Attenuation of Visual Persistence," _British Journal of Psychology,_
Vol. 63, No. 2, May, 1972, pp. 243-254.

KEY WORDS:

Man	Setting	Relational Concepts	Outcomes
Students	*Laboratory*	*Trace brightness, spatial separation, field illumination*	*Visual persistence*

Several studies have examined the interaction between the nature of
visual stimulation, and the persistence with which that stimulus will be
viewed. In general, persistence has been found to be greater for short
exposures than for long exposures, it varies according to the brightness
of the stimulus or surrounding environment, and abruptly terminates by
the occurrence of a visual noise mask. This study was designed to apply
certain methodological improvements not utilized in previous investigations in order to measure visual persistence (VP) more directly.

Three separate experiments were implemented using different
undergraduate or postgraduate students for each one. In the first two
experiments a cathode-ray oscilloscope (CRO) produced the stimuli. When
the CRO was set in motion the resulting visual field appeared to be a
number of intersecting lines. By changing the rotational speed of the
CRO it was possible to reduce or increase the number of lines which
could be seen.

In experiment one, seven undergraduate students who had normal or
corrected to normal vision were exposed to 12 conditions, including two
levels of trace brightness, two levels of field illumination (2.0
ft.-lm. and 0.02 ft.0 -lm.), and three rotational speeds of the CRO
(0.4, 0.8, and 1.0 rev./sec.). The sweep speed was gradually decreased
so that the number of lines would increase. Each subject was told that
the number of lines would gradually increase from two to five. At the
point when the subject saw one more line than previously, he was
instructed to respond. At that time the experimenter marked down the
corresponding sweep speed. Results from this segment of the study
revealed that although the changes in VP were comparable with earlier
studies, the absolute magnitudes of VP were not. That is, an increase
in stimulus brightness and surround illumination resulted in a decrease
in persistence, but the magnitude was smaller than might have been
expected.

The second experiment examined whether "varying the spatial separation of successive stimuli on the retina by a method other than changing
the speed of rotation would have a comparable effect on VP." The CRO
was manipulated so that changes in the viewing distance would be simulated. Eight subjects were involved in eight trials similar to the
first experiment. Results did not reveal the predicted hypothesis that
increased viewing distance would lead to a decrease in VP. In fact,
there was a small but insignificant increase.

The third experiment was designed to test the hypothesis that
increasing spatial separation would result in progressively greater

values of VP. A stroboscope was used to allow different rates of stimuli flashes. Three postgraduate students took part in the procedure, which was similar to the first experiment. The results indicated that VP is an increasing function of the spatial separation of successive stimuli.

The authors concluded that VP varies considerably in response to situational demands. "It seems that VP is greater in conditions that are visually unfavorable (dim illumination, short exposures) and is reduced in favorable conditions."

DRINKWATER, B.L., P.B. RAVEN, S.M. HORVATH, J.A. GLINER,   63
R.O. RUHLING, N.W. BOLDUAN, AND S. TAGUCHI

"Air Pollution, Exercise, and Heat Stress," *Archives of Environmental Health,*   Vol. 28, No. 4, April, 1975, pp. 177-181.

KEY WORDS:

Man	Setting	Relational Concepts	Outcomes
Healthy young males	*Laboratory*	*Air pollution, aerobic power*	*Revised air pollution standards*

Convincing evidence exists that acute levels of air pollution are hazardous for the elderly and those people with respiratory and cardiovascular problems. However, little is known regarding the possible physiological difficulties imposed upon healthy persons who live and work in urban environments. This laboratory study was designed to test twenty healthy young men under four different air conditions in order to obtain an indication of potential relationships between air pollution and aerobic power. The four air mixtures used were: 1) filtered air (FA); 2) 50 ppm carbon monoxide (CO); 3) 0.27 ppm peroxyacetyl nitrate (PAN); and 4) CO plus PAN (PANCO). Temperature was kept constant at 35C under all test conditions.

Each subject initially spent a five minute rest period during which time the air mixture in the environmental chamber was breathed. Immediately following this rest period a pace of 94 meters/minute on a treadmill was set. The grade of the treadmill was increased 1% each minute during the walk. In order to measure maximum aerobic power the subject was asked to walk until he indicated that he could no longer continue. During the testing, rectal, skin, and room temperatures were taken twice a minute and an electrocardiogram was in constant use. After a 15 minute recovery period, the subject completed a 33 item questionnaire in order to get a subjective indication of the physical and mental state he had experienced.

Maximal aerobic power was found to be unaffected by exposure to CO, PAN, or PANCO. It was discovered, however, that there was a decrease in the work time for smokers in the CO environment. Thus, the authors concluded that "since the concentrations of CO represented a first-stage alert level for the Los Angeles area, it appears that the air pollution standard for this pollutant is set at a realistic level for healthy young men doing strenuous work for a relatively short period of time."

They also emphasized that this standard does not necessarily hold true
for all types of people, nor does it represent the possible outcomes of
exposure to the environment for longer periods of time.

DUNCAN, J.S. JR.   64

"Landscape Taste as a Symbol of Group Identity," *Geographical Review*,
Vol. 63, No. 3, July, 1973, pp. 334-355.

KEY WORDS:

Man	Setting	Relational Concepts	Outcomes
Suburbanites	*Residential land-scapes*	*Landscape configuration*	*Social stratification*

The ways in which home owners configure their environment is one
way of determining their self-perceptions and the image they wish to
project to the surrounding community. Although this hypothesis has been
used in the past to distinguish different levels of social groups, the
author of this study has suggested that the subtle variations of landscape taste of two groups that are nearly identical in socio-economic
status are significant indicators of group identity.

A physical inventory of 1139 residential landscapes in Bedford
Village, New York, was performed during a three month period in 1971.
The landscapes were classified by age, size, relation to entry road,
degree of seclusion from surrounding residences, and degree of vegetation cover. The residences were classified as either Alpha to denote
the older, more established manors of the original village or Beta as
characterized by the more recent additions to the village in tract lots
and subdivisions. Although both landscapes were well maintained and had
high realty values, the Alpha types tended to be more secluded geograph-
ically and had more defined boundary fences and hedges. The Beta types,
on the other hand, were clustered in a more communal manner and had more
access to public areas.

The author matched the Alpha and Beta residences with membership in
specific social and religious registers. There was a high correlation
(84%) between a Protestant affiliation and residence in the Alpha area
and between the Catholic Church and the Beta area (83%). The same correlations were found in other registers such as membership in the fire
department, historical society, and country club. The author argued
that landscape configurations were used in these two environments to
advertise a specific social grouping and to reinforce a self-image.

EDNEY, J.J.   65

"Place and Space: The Effects of Experience with a Physical Locale,"
Journal of Experimental Social Psychology, Vol. 8, January, 1972, pp.
124-135.

KEY WORDS:

Man	Setting	Relational Concepts	Outcomes
Students	*Room*	*Territoriality*	*Interpersonal distance*

Territoriality was discussed in terms of effects on both animal
species and human populations. Casual observation of the social behaviors of humans suggested that a relationship exists between personal
space needs and the use of physical environments. Dominance hierarchy,
distinctive positioning, social isolation, and personality compatability
factors have all been associated with human territoriality. The literature suggests that two factors mediate the impact of physical space on
behavior: 1) a person's prior experience with a place; and 2) the anticipation of future experience with a place.

This study sought to delineate changes in specific behaviors and
reactions that occurred as a function of the association of a person
with a physical space over controlled time periods. Past and anticipated future experiences were measured. Interpersonal distance viewed
as a social index, was assessed simultanteously with perceptions of the
size of the physical surroundings and as a measure of space usage.
Ninety-six college undergraduates were each allowed a controlled experience in a room. A 2 x 2 x 2 design was used in which half the subjects
were given prior exposure to the room, half were induced to anticipate
future experience with it, and half were joined there by a stranger.
Questionnaires were used: 1) to elicit person's ratings of the room on
a number of dimensions; 2) to have subjects select one of a number of
alternative changes to the room; 3) to allow subjects to complete sentences that elicited person's emotive responses to the room; and 4) to
ask subjects to conjure a number of interior schemes for the room.

In the second half of the hour long test, each subject responded to
a second group of items and tasks comprising the dependent measures. An
intruder systematically entered the testing setting. Attitudinal, perceptual, and behavioral measures were employed to elicit additional
responses.

Results showed that choice of interpersonal distance was reduced
both by prior and expected experience with the physical setting.
Exposure to the room significantly affects perception of its size.
Exposure negatively affected subject's evaluation of the place.
Anticipation of future experience with the place interacted with the
presence or absence of another person there. This influenced subjects'
moods and actual behaviors in the space. The author concluded by
discussing relations between these results and current knowledge of
territorial behavior.

"Crowding and Human Sexual Behavior," *Social Forces,* Vol. 55, No. 3, March, 1977, pp. 791-808.

KEY WORDS:

Man	Setting	Relational Concepts	Outcomes
Families	*Urban*	*Crowding*	*Sexual behavior*

The data for this study came from a stratified probability sample of Toronto families living in areas selected both for the range of compressed neighborhood conditions and for their potential in yielding a large number of families in which the number of people exceeded the number of rooms they live in. The population sampled (560 households) had one or more children, the female being under 45 yrs of age, and had lived in their present location for at least three months. One or both parents were interviewed by physicians for two hours and given medical exams, which included a series of questions concerning sexual behavior. In all, 294 women and 213 men were interviewed and medically examined.

Thirty measures, used to tap various aspects of crowding were incorporated with the study and three aspects of the married pair's sexual relations with each other were explored as dependent variables. Multiple regression techniques were used to assess the effect of crowding on the various types of sexual behavior considered and the results indicated that crowding has an effect only on reports that lack of privacy interferes with marital intercourse. Even then, it was not the actual crowded conditions but the feeling that the household is crowded that influences such behavior. The difference for men and women was significant with the males reporting a much lower rate of intercourse and even cessation of intercourse for periods of time due to crowded households.

The feeling of stress served to intensify the social affects of crowding with men reporting more frequent intercourse and more extramarital relations in congested neighborhoods if high stress was also a factor. Early experience with crowding in one's lifetime combined with stress at present tended to produce more females who are sexually involved with someone other than their spouses. On the basis of the findings, human sexual behavior (marital, extramarital, homosexual or incestuous) appeared to be appreciably influenced by crowding in only highly selective circumstances and, even then, the effects were quite different from effects noted in research done with animals and crowding.

EISEMON, T.   67

"Simulations and Requirements for Citizen Participation in Public Housing: The Truax Technique," *Environment and Behavior*, Vol. 7, No. 1, March, 1975, pp. 99-123.

KEY WORDS:

Man	Setting	Relational Concepts	Outcomes
Residents	*Public housing*	*Simulation techniques*	*Preferences*

A technique was developed and implemented for eliciting citizen involvement in the design of low-cost public housing. A simulation device was created to enable citizens to express their housing preferences visually rather than verbally. The simulation was field tested in 1971 with residents of the Truax housing project in Madison, Wisconsin and was later revised. The need for increased citizen involvement in the urban public decision making arena was discussed. Objectives were to develop a workable strategy which: 1) explicitly involved citizens in decision making roles; 2) was suitable for use with community groups and could also be effective in the entire urban area; 3) minimized the reliance of citizens on professional expertise; and 4) provided public officials with useful data.

The Truax technique encouraged citizens to articulate their housing needs through a gaming simulation device requiring them to annually simulate the construction of a housing unit reflecting needs and desires that are simultanteously commensurate with specific financial realities. The technique was viewed as suitable for subject groups affected by public housing programs to whom more abstract data elicitation techniques might be intimidating.

Twenty-one residents of the housing development participated in the initial field experiment. After the initial gaming device was administered, these preferences were linked to the actual conditions of the housing development. Also, proposals were generated by architects which were then presented to residents. The raised simulation device contained six tasks required of participants. The most useful result of the second field experiment was the successful development of alternative housing plans. Differences in floor plan preferences were detected between males and females. The gaming technique was found to meet the initial objectives of the project. Additionally, it was concluded that architects can become sensitized to users needs through their participation in this type of housing preference analysis.

EISENBERG, T.A. AND R.L. McGINTY   68

"On Spatial Visualization in College Students," *Journal of Psychology,* Vol. 95, January, 1977, pp. 99-104.

KEY WORDS:

Man	Setting	Relational Concepts	Outcomes
Students	*Laboratory*	*Career Orientation*	*Spatial Visualization*

It has been assumed in past research that the ability to visualize and conceptualize in three-dimensional space is an important factor when people choose a career field. Architects and engineers, for example, are considered to have innate visual talents in order to succeed in their professions. This particular study tested the hypotheses that persons with specific career goals have predicatable visual abilities and that there are significant differences in spatial visualization between the sexes of similar career groups.

The study used four different groups of student populations in order to capture a range of career orientations. Seventy-two students were enrolled in a business statistics course (B), 37 in an advanced calculus course (C), 58 in a remedial math course (R), and 56 in a course designed for future elementary school teachers. Each group was administered a 20-minute, 27 item multiple-choice exam incorporating four types of spatial reasoning. The exam contained such items as two and three dimensional geometric puzzles, spatial word sequences, and figural analogies. Sex variations within and among groups were taken into account during data analysis and test results were compared with each student's ACT composite score using an analysis of covariance technique.

It was found that in three of the four test areas, the C group scored significantly higher than any of the other three populations, ($r = .05$). An unexpected result of the study was that females in the normally male-dominated calculus and business classes scored significantly ($r = .05$) higher than did males. The same was true for males in the teacher group. In general, the experiment reinforced the hypothesis that people with different career goals significantly differ in spatial aptitudes and that differences among the sexes in spatial reasoning is also present.

ELSON, M.J.   69

"Activity Spaces and Recreation Spatial Behavior," *Town Planning Review,* Vol. 47, No. 3, July, 1976, pp. 241-255.

KEY WORDS:

Man	Setting	Relational Concepts	Outcomes
Car Owners	England	Recreational areas, preference	Spatial search behavior

In an attempt to learn more concerning the use of activity space in the county of Sussex, England, a random sampling of car owners was carried on in Lewes, the county seat, which had a population in the early 1970's of approximately 14,000. Using the Electoral Register, over 1000 persons were contacted and asked if they owned a car. Those who were car owners were given a questionnaire to fill out. A total of 696 valid responses were used for the final analysis and the response rate was 65.6%.

Relatively more responses came from the more affluent and older population of Lewes. Questions on the use of recreation and activity space in the county showed that the coastline provided the major attraction for these car owners, even though in the East of this country, where Lewes is located, the coastline was far less accessible than the forest areas which had been developed as recreation/activity centers.

The hypothesis tested in this survey was that the process of spatial search (the development of activity spaces through a period of time) continually expands. Results showed that the total of activity locations visited even once seemed to increase rapidly until 4 years after an individual or family had settled in Lewes, when the visits to new areas leveled off. In the 5-10 year period of residency in the area, families and individuals gradually increased the number of recreation areas visited until an averge of 10 locations was reached.

A clear picture emerged of the spatial search behavior of the new arrival in the town of Lewes: 1) search made first along coastal regions for areas which would provide recreation activities, 2) search made through the "downland" (southern, hilly) area, and 3) search made inland (East Sussex) in forested area. More data must be collected, especially concerning the less affluent families in the area, before a model can be developed to show that there are ways to determine the probability of visitations to locations set aside for recreation purposes in order to assure maximum usage.

EVERETT, M.D.   70

"Roadside Air Pollution Hazards in Recreational Land Use Planning," _American Institute of Planners_, Vol. 40, No. 2, March, 1974, pp. 83-88.

KEY WORDS:

		Relational	
Man	Setting	Concepts	Outcomes
Recrea-	_Recreation_	_Recreation_	_Facility usage_
tionists	_areas_	_pollution_	

There is strong scientific evidence that living and working in highly polluted air can significantly increase the change of developing chronic respiratory disease. Many recreation activities take place in areas of heavy traffic. Today, more cyclists than ever are riding in or along busy roadways, joggers have been common along the roads for years, and tennis courts, tracks, and gymnasiums are often initially located in thoroughfares or on streets which subsequently are "improved" and become major arteries. This study, which summarized the available scientific literature on levels of roadside air pollution and the health impact of exercising in polluted air, addressed the question whether recreation carried out in these areas exposed the participants to high levels of air pollution.

The literature clearly indicated that recreation facilities and activities near heavily traveled roads expose participants to levels of air pollution considerably higher than general ambient levels in the same urban area. Evidence existed that air pollution is harmful to health, that roadside air pollution is high relative to general ambient air pollution, and that the additional stress of exercising may exacerbate the negative health effects of air pollution.

The author offers four general suggestions for the site selection of recreational land uses. Gymnasiums, tracks, tennis courts, pools, playgrounds and other facilities for physically active recreation should be constructed at least one block away from major arteries. Environmental impact statements on road construction and improvements around existing recreation facilities should consider air quality standards.

Although accessibility is a major consideration for spectators of professional athletic events, facilities for amateur events should stress
air quality. Regional and larger urban parks should limit automobile
access to some areas and utilize hiking, cycling, and canoeing routes to
provide access to semiprimitive camp grounds and recreation areas. The
author further suggested that the crude indices of pollution levels
along various types of roads that should be used when evaluating
recreation areas could also be employed for all types of land use decisions, including residential and employment.

EVERITT, J.C.   71

"Community and Propinquity in a City," _Annuals of the Association of_
American Geographers, Vol. 66, No. 1, March, 1976, pp. 104-116.

KEY WORDS:

Man	Setting	Relational Concepts	Outcomes
Husbands and wives	Urban area	Propinquity; cognitive mapping	Site recalcitrance; behavior patterns

An empirical study of selected forms of behavior of husbands and
wives in West Los Angeles demonstrated significant distance and directional biases for both groups. The study goals were to recognize spatial and physical constraints, and the ability of subjects to accurately
map them using cognitive mapping procedures. The concepts "community"
and "neighborhood" were differentiated. The author cited that although
both concepts have been used to evaluate the pattern and depth of social
relationships, community is viewed here as concentrating on human behavior, whereas neighborhood is more spatial. The urban individual was
conceptualized as belonging to two basic kinds of communities; those
with, and those without propinquity. The restriction of information in
a community influences the individual's image. This restriction was
termed site recalcitrance. The concept of community without propinquity
suggests that individuals interact within a large area of the city, and
that site recalcitrance places parameters upon this interaction. Three
hypotheses were tested: a) the activities of individuals will be
nonpropinquitous and biased by site recalcitrance; b) the activities of
wives were more propinquitous than those of their husbands; and c) the
city dwellers can indicate areas of significance by cognitive mapping
technique.

The sample group was comprised of 65 married households. Various
activity patterns of subjects' lifestyles were studied to test hypotheses A and B. The third hypotheses focused on the relationship between
perception and behavior patterns. The principal activities studied
were: a) trips to the workplace; b) homes of friends; and c) formal
organization memberships. Propinquity was defined as the "nearness of
place".

Results indicated that both distance and direction are important
considerations in the analysis of individuals within the urban area. A
decrease in total activities was detected at a distance between four and

six miles from the center of the study area. The tendency for activities to have a directional bias toward the north and east was also significant. The nonpropinquitous behavior of the study group confirmed hypotheses A and B. Site recalcitrance was a result of physical and sociocultural factors acting together to restrict the subjects image of the city, and thus behavior within the setting.

This environmental restriction was evident regarding distance, where role differences between husbands and wives resulted in differing levels of environmental apprehension. This occurred despite identical opportunity sets available to the two groups. The structure of the city was found to have a major influence on subjects activities within it.

Spatial patterns resulting from the cognitive mapping techniques showed considerable congruence with other data collected. Physical distance was influential in the selection of places frequented most often. The differences of behavior patterns between husbands and wives viewed as a result of their different roles in urban society. The degree of each group's identification with their local area was deemed worthy of more detailed study.

FANGER, P.O.   72

"Thermal Environments Preferred by Man," *Build International*, January-February, 1973, Vol. 6, No. 1, pp. 127-141.

KEY WORDS:

Man	Setting	Relational Concepts	Outcomes
Students	*Laboratory, Denmark*	*Thermal factors, age, adaptation, sex, circadian rhythm*	*Thermal comfort, thermal preference*

The author reviewed the existing knowledge of the conditions for thermal comfort for men and women and discussed the interction between physiological comfort conditions and the air temperature, mean radiant temperature, air velocity, humidity, clothing, and activity type. In addition, preliminary research data were presented for thermal comfort in terms of age, adaptation, sex, and circadian rhythm.

Thermal comfort is defined by the author as "that condition of mind which expresses satisfaction with the thermal environment." Generally, this concept indicated that the individual does not desire to be in a warmer or cooler environment. However, since people are not alike, it will not be possible to satisfy everyone at the same time. Consequently, for the purposes of this study, speaking of creating the optimal thermal comfort means determining that condition in which the highest possible percentage of the group was thermally comfortable.

Earlier studies by Yaglou (1927), Gagge, et al (1937), Winslow, et al (1937), Dubois, et al (1952), and Nielsen (1947), have revealed a correlation between thermal sensation and skin temperature, regardless of whether the subjects were clothed or not. Thus, it was generally accepted that the physiological conditions for comfort revolved around maintaining a mean skin temperature of 33-34° C° and not allowing sweating to occur. A 1967 study by the author confirmed these findings, but only for those subjects performing sedentary activities. At higher activity levels, a lower mean skin temperature and a preference for sweating were observed. In more recent experiments, Wyon, et al (1971) revealed that substantial savings of the temperature were allowable if they take place rapidly, provided that the mean values of the physiological variables are kept at the comfort values.

Fanger's assumption was that it is the combined thermal effect of all physical factors which is of importance for man's thermal state and comfort. Consequently, it was impossible to consider the effect of any of the physical factors influencing thermal comfort independently since the effect of each of them depends on the level of the other factors. Several studies were undertaken in a controlled environment by the author for a total of 1300 subjects. The procedure was to dress each individual in a standard uniform (0.6 clo), expose them to varying levels of temperature and humidity, take physiological measurements at standard predetermined times, and have the subjects vote every half hour on the Bedford 7-point scale.

Results can be summarized as follows: 1) no thermal preference variations were found between 129 Danish college-aged subjects and 128 elderly subjects; 2) no differences between the sexes of the elderly group were apparent; 3) it was not found to be the case that man cannot adapt to prefer warmer or colder environments, which implies that the same comfort conditions can be applied to buildings in the tropic as in temperate or polar climates; 4) men and women seemed to prefer almost the same thermal environments (women slightly lower); 5) there was no difference between the comfort conditions in winter and in summer; and 6) there were only insignificantly preferred ambient temperature fluctations throughout the day.

FELTON, B. AND E. KAHANA   73

"Adjustment and Situationally-Bound Locus of Control Among Institutionalized Aged," *Journal of Gerontology,* Vol. 29, No. 3, May, 1974, pp. 295-301.

KEY WORDS:

Man	Setting	Relational Concepts	Outcomes
Elderly	*Institution*	*Locus of control*	*Positive adjustment*

Prior to this study, perceived locus of control had not been studied among the institutionalized elderly. Previous research was cited concerning locus of control as a possible explanation for certain life satisfaction levels and personality attributes of elderly persons. In this study locus of control was viewed in situational contexts (i.e., contexts which reflect the potential for real control over the environment in the institutional setting). The relationship between locus of control and environmental adjustment among the elderly was examined using residents' solutions to nine hypothetical problems considered typical of congregate living settings as the measure of perceived control. The sample consisted of 124 residents of three retirement homes. Subjects ranged in age from 55 to 97. Seventy-four percent of the subjects were female.

Interviews were conducted on an individual basis. A pre-test measured the effectiveness of the testing procedure prior to the formal interviewing sequences. Hypothetical situations represented the problems of monotony, privacy, conformity, emotional expression, activity, environmental ambiguity, affective intensity, motor control, and autonomy. Respondents' solutions to the problems comprised the local of control measure. Responses were coded into the final categories of self, staff, and others. Adjustment was measured via: 1) staff-rated satisfaction scales, where staff members rated residents on a 1 to 10 scale; 2) staff-rated adjustment scales, where staff members rated residents on the same scale regarding adjustment; 3) Lawton's morale scale, as a 22 item device; and 4) self-rated life satisfaction ratings, where each subject placed himself on a 1 to 10 point ladder in terms of self-perceived life satisfaction.

Results of an analysis of variance indicated that for four of the nine problematic situations, perceived locus of control related positively to adjustment. These included problems related to environmental ambiguity, autonomy, emotional expression, and privacy. In five of the six significant relationships, externally rather than internally perceived control was found to relate to postive adjustment. Perceived locus of control appeared to possess a different meaning in institutional settings where persons "external to the individual may function as intermediaries between the powerless self and a rigid institutional environment." A series of possible explanations for the results were presented. In most cases, individuals who perceived staff and others as the locus of control were better adjusted to the built environment. It was concluded that further specification of the situational context of control and relationships to long term control behaviors in the elderly might provide important clues to their process of environmental adjustment.

FIEDLER, F.E. AND J. FIEDLER   74

"Port Noise Complaints: Verbal and Behavioral Reactions to Airport-Related Noise," *Journal of Applied Psychology,* Vol. 60, No. 4, August, 1975, pp. 498-506.

KEY WORDS:

Man	Setting	Relational Concepts	Outcomes
Neighborhood residents	Residential areas near airports	Noise	Attitudes/observed behaviors/tolerance

This study compared the verbal and behavioral reactions of residents living in high, medium and low-noise zones surrounding a major airport, as well as in communities out of the airport's noise range.

The authors capsulized the body of literature related to this subject and stress that jet noise is an ever-present feature of the environment near a major airport, and that there is considerable controversy over its effect on the health and well-being of people who live under these conditions. Major goals were to prove the value of using two different types of control areas outside the airport's noise zone, and a collection of nonreactive (unobtrusive) measures that were indicative of the residents' life-style in response to living in areas that objectively differed in quality and quantity of prevailing noise exposure levels.

The investigation was conducted at the Seattle-Tacoma International Airport. Interviews were conducted with subjects in the Airport (N=302), Control (N=98), and County (N=316) groups. Information that was elicited included relevant demographic characteristics, educational backgrounds, occupational-income data, and individual attitudes concerning ecological and community problems as well as reactions to airport and airport related noise. This interview lasted from 30 to 45 minutes. Behavioral inventories were also made of outdoor activity patterns in the three zones under study.

Results showed that the effects of noise were highly complex. While the proportion of those reported to be bothered by noise was correlated with objectively measured noise levels, the intensity and perceived source of the noises were unrelated to reported psychological and physical symptoms, length of residence in the area, and/or trace measures indicating recreational use of outdoor areas. It was also shown that even high noise levels as those generated by approaching and landing jet aircraft may have little effect on changing the behavior and life-styles of residents. The authors concluded that the marked contrast between noise related attitudes and behaviors needs to be carefully considered in drawing conclusions about the "real" effects of noise pollution in the environment.

FITCH, J.M., J. TEMPLER, AND P. CORCORAN   75

"The Dimensions of Stairs," *Scientific American,* Vol. 231, No. 4, October, 1974, pp. 82-90.

KEY WORDS:

Man	Setting	Relational Concepts	Outcomes
Human	*Stairways*	*Stairs, use of/ design of*	*Safety on stairs*

Architects have not yet evolved a holistic theory of human deployment in space and this fact becomes apparent when the design of stairways is discussed. The design of stairs imposes a certain gait on humans and the restraints in that gait are both physiological (in terms of energy required to lift or lower the body from one point to another) and cultural (as dictated by the customs and costumes of the culture).

This study was prompted by the many accidents reported on the outside steps at Lincoln Center in New York City which are especially difficult to see due to shallow risers and the color of the material used.

Using replicas of four actual stairways, sixteen men and women, wearing
ordinary street wear and shoes of their own choice, were photographed
going up and down these stairways. Also used in the study was a mechanical stairway which operated as a treadmill to test the range of riser-
tread combinations. This was done at the metabolic research laboratory
of Helen Hayes Hospital in West Haverstraw, New York.

Results indicated that risers from 6.3 to 8.9 in. and treads from
7.7 to 14.2 in. produce stairs that are safer to climb, but when
descending, the treads which result in the fewest missteps at any speed
are at least 12.3 in. deep. Treads less than 9 in. deep are associated
with the greatest number of missteps regardless of the height of the
riser or speed of ambulation. If stairs are to be adequate for the
taller, larger populations, the tread must be at least 11.8 in. deep.

Using 689 observations of subjects on the treadmill stairway, an
equation was formulated to predict the energy demand of different stair
ascents. Again, findings suggest a stairway designed with risers from 4
to 7 in. coupled with treads 11-14 in. will result in low rates of
energy expended at normal ambulatory speeds and a low rate of missteps.

FLOCK, H.R. AND K. NOGUCHI   76

"Brightness Functions for a Complex Field with Changing Illumination and
Background," *Canadian Journal of Psychology,* Vol. 27, No. 1, March,
1973, pp. 16-38.

KEY WORDS:

Man	Setting	Relational Concepts	Outcomes
Students	Laboratory	Brightness contrasts	Brightness percep- tion

Earlier studies involving the effects of illumination changes on
the perception of various background fields have indicated that under
certain conditions when illuminance is increased, brightness judgments
decrease rather than increase. The purpose of this research was to
diminish possible ambiguities of previous experimental design techniques
and to build on the knowledge gained from the earlier studies.

Ten undergraduate students who were unaware of the purposes of the
experiment, and who had at least 20/20 uncorrected vision in each eye
took part in the study. Each subject was seated in front of a viewing
tunnel that was enclosed in a black wooden box and allowed the left eye
to view a cross-shaped configuration of seven neutral grays mounted on a
gray background. An identical arrangement was constructed for the right
eye, but in order to alter the illumination of the right eye display
field, seven different combinations of lenses could be placed in front
of the viewing lens. The authors emphasized that "although these
arrangements of the apparatus produced a dichoptic viewing situation,
the apparatus was arranged so that the left-eye field and the right-eye
field could not be imaged on the fovea simultanteously." A total of 168
combinations was presented to each subject in a random order. During
the three hour experimental session subjects were asked to indicate the

apparent brightness of the the test field by assigning an appropriate number to it.

The results indicated that, on the average, a white background produces lower brightness judgments than does a black background. Furthermore, as the background was shifted from black to gray, it was discovered that the brightness of the black background increases at each level of illumination, instead of the contrasting related literature findings which indicated that it decreases. The authors hypothezized that the reversal of findings was a result of the fact that most other studies were "based on a simple stimulus array, involving a homogenous focal region at some level of luminance and a second adjacent region at a different level of luminance," whereas, "in the present experiment the stimulus arrays were less simple, involving eight regions of different luminances." They concluded that since the same types of findings were also obtained in a similar 1970 study by Flock and Noguchi, the reversals were probably reliable.

FLOCK, H.R., K. NOGUCHI, AND P.M. MUTER   77

"Lightness Changes in a Complex Field with Changing Illumination and Background," _Canadian Journal of Psychology_, Vol. 28, No. 4, December, 1974, pp. 446-467.

KEY WORDS:

Man	Setting	Relational Concepts	Outcomes
College students	Laboratory	Lightness	Lightness perception

Much confusion has existed in the literature regarding simultaneous brightness contrast due to a misunderstanding of the differences between the concepts of brightness and lightness. "If judgement of apparent brightness were the task, and if the subject were expected to act like a photometer as in a defining brightness experiment, then he must ignore as best he can the surround region and respond only to the light intensity of the focal region. If the subject's task is lightness judgement, he will be expected to act in a very different manner and register the contrast-relationship across the edge between surround region and focal region, on which the achromatic appearance depends." The authors believed it is a misunderstanding of the differences between these two concepts which explains the contradictory findings of earlier lightness studies. Consequently, it was the purpose of this experiment to ensure that participants understood the concepts, so that results would be more accurate, and uncertainties regarding lightness functions would be greatly diminished.

Ten undergraduate students who were unaware of the purposes of the experiment, and who had at least 20/20 uncorrected vision in each eye took part in the study. The apparatus used was identical to the one described in Flock and Noguchi, 1973. In this experiment three separate scales were used: one for black backgrounds, one for grays, and one for whites. Use of the three scales in the experiment was preceded by extensive training so that ambiguities could be avoided, and a more constant set of responses achieved.

It was found that illumination did result in systematic overall effects on the lightness responses in most of the 24 test fields examined. In addition, when the lightness functions of gray- and black-appearing test fields were examined at upper levels of illumination, they were discovered to be flat. Over lower levels of illumination, three-fourths were incremental, and the remaining quarter were flat. Thus, although these findings tended to contradict the results of other lightness studies, the authors contended that their findings are consistent with the major theories of illumination. They further hypothezied that the discrepancies resulted because of ambiguities of what was being measured and by the methodologies which were utilized.

FOX, D.J.   78

"Patterns of Morbidity and Mortality in Mexico City," *Geographical Review,* Vol. 62, No. 2, April, 1972, pp. 151-185.

KEY WORDS:

Man	Setting	Relational Concepts	Outcomes
Mexican	*Housing*	*Sanitation*	*Mortality*

This study was an attempt to map the relationships between living environments and fatal diseases. Through various statistical techniques, the author has shown that specific types of housing arrangements can adversely affect the health of their occupants and that improvement in the standard of housing over the past century has dramatically reduced the overall death rate.

The study was based on death records for a three month period in 1965 that showed the causes of mortality in fifteen selected housing districts of Mexico City. Housing types were established through the use of an official Mexican survey that listed specific environmental conditions in the fifteen areas in terms of interior sewage connections, sanitary water connections, ventilation, and other physical characteristics. The death records indicated the causes of death and were sufficiently accurate to pinpoint the decedents' domiciles.

There was a strong correlation between infant deaths and the lack of interior toilet and kitchen facilities. Adult deaths with a significant relationship to unsatisfactory housing conditions were due to infective and parasitic diseases, although most environmental causes of adult deaths were not nearly as conclusive as those of infant mortality. Poor ventilation and death due to respiratory infections showed a high correlation, while cancer and cirrhosis of the liver were not tied to housing conditions at all. Overall, the study showed that the three areas defined by the housing survey to be most inferior had the highest infant death rate and a significantly higher adult mortality rate. The author concluded that improvements since 1920 in the environmental conditions in Mexico City had lowered the death rate from 40 to 10 per 1000 and that continued improvements would make a strong impact on the reduction of infant deaths.

"Problem of the Old and the Cold," _British Medical Journal_, January 6, 1973, pp. 21-24.

KEY WORDS:

Man	Setting	Relational Concepts	Outcomes
Elderly	_Residence_	_Thermal comfort_ _Age_	_Thermal awareness_

Much uncertainty exists regarding the importance of endogenous and exogenous factors in the causation of hypothermia (body temperature less than 35° C) in the elderly. This study was designed to examine the living conditions and environmental temperature of elderly people living in their homes.

One hundred subjects were chosen from 300 people in 1971 who were 65 years of age and older and who were receiving the "meals-on-wheels" service in Portsmouth, England. The age range of the subjects was 65-91, with the average age being 77 years.

In order to discover how body temperatures correlate with factors in the "life situation," data was collected between January 18 and March 10, 1971 in the form of seven measures: 1) mouth temperature, 2) hand temperature, 3) urine temperature, 4) room and air temperature, 5) thermal body comfort votes by subjects, 6) thermal hand comfort votes, and 7) actual weather conditions.

Although the results did not reveal a serious case of hypothermia, it was discovered that in 81.3% of the cases the rooms were below the Parker Morris minimum comfort limit of 18.3 C. Furthermore, there were strong indications of a relation between those with "low" body temperatures and inadequacy of heating arrangements, the lack of basic amenities, and the subject's response to certain aspects of the "life situation." The lower temperature group tended to be living in the coldest dwellings with the greatest deficiencies in basic amenities which would help to keep them warm. At the same time, many of them did not feel that they were cold or even shivered, and they did not think the heating level was inadequate. The authors concluded that "these apparently lower warmth and cold thresholds in some old people, combined with a low environmental temperature in their homes and a reduced safety margin of heat to lose before hypthermia will supervene, would seem to put them especially 'at risk'."

FOX, R.H., P.M. WOODWARD, A.N. EXTON-SMITH, M.E. GREEN,   80
D.V. DONNISON, AND M.H. WICKS

"Body Temperature in the Elderly: A National Study of Physiological,
Social, and Environmental Conditions," *British Medical Journal*,
January 27, 1973, pp. 200-206.

KEY WORDS:

Man	Setting	Relational Concepts	Outcomes
Elderly	Residence	Thermal comfort Age	Hypthermia suscept-ibility, thermal awareness

Two large-scale surveys of body temperatures in elderly people
living at home in England were carried out during January, February, and
March 1972. There were three major aims in the study. The first was to
ascertain the incidence of low body temperatures in the elderly at home
during three months using recently developed physiological techniques.
Secondly, to relate the body temperatures to the environmental temperatures and living conditions. The third was to identify categories
of elderly people particularly at risk of developing hypothermia (body
temperatures less than 35º C) by investigating the relationships between
body temperatures, the individual's awareness of warmth, and socioeconomic status.

The sample consisted of over 100 people 65 years of age and over
located in 100 different sample points. The major difference between
this survey and an earlier study by Fox is that in this study body temperatures were taken in the morning and the afternoon, rather than
being combined together. This difference proved to be crucial since the
morning measurements revealed hypothermia to be of a greater magnitude
than was indicated by the 1971 survey of elderly living in Portsmouth,
England. When deep body temperature was taken in the morning, 10% of
the sample was found to have a temperature less than 35º C. Yet, by
evening, all of the temperatures of these individuals had risen during
the day so that none of them appeared to be experiencing hypothermia.
Thus, the problem was emphasized to a greater degree by the more recent
study. "These people are 'at risk' not only because they are less successful in conserving body heat, as shown by their inability to maintain
an adequate core/shell temperature gradient, but also because they have
a proportionately lower body heat content. This study underlies the
lack of awareness of cold by some old people." One suggestion offered
by the authors as a result of these findings is that those elderly who
are "at risk" should be provided with a waterproof, electrically safe,
low-wattage electric blanket to combat the lower body temperatures
experienced during the late-night and early-morning hours.

"Human Response to Buffeting in an All-Terrain Vehicle," *Aviation Space and Environmental Medicine,* Vol. 47, No. 1, January, 1976, pp. 9-16.

KEY WORDS:

Man	Setting	Relational Concepts	Outcomes
Males	*All terrain vehicle*	*Buffeting/human response*	*Effect on body joints*

All terrain vehicles (ATV), both wheeled and tracked, are used by many persons for recreational, professional, and/or military purposes. Considerable attention has been paid in the past few years to the dangers of crash injury and death during the operation of these vehicles, both on and off the road, and resulting publicity has led to legislation at different levels of government. Much less concern has been given to the dangers of chronic bone and joint pathology from exposure to the vibration and buffeting experienced in operating these vehicles. The purpose of this study was to determine under field conditions the acceleration loads encountered in operating small, personal, multi-wheeled ATVs and snowmobiles.

For the experiment, the initial vehicle selected was a popular, six-wheel, low-pressure-tire ATV, designed primarily for an operator and one passenger. Each subject was fitted with a helmet to which was bolted a mount to receive a transmitter to send electronic impulses to a four-channel FM instrumentation recorder. The track was rectangular and heavily plowed across the direction of travel. Eleven paid, male volunteers were selected from among a student population ranging in age from 18-23. Each subject was instructed to drive the vehicle at maximum speed compatible with safe handling.

An analysis indicating acceleration load, peaks and frequencies was calculated and the particulars of this analysis are presented in the study. The experience, subjectively determined from the debriefing of subjects, was exhilarating, fatiguing, and occasionally frightening, even for a three minute run under the moderately severe circumstances of the test track. This study provided an illustration of the nature and severity of the forces encountered, and an indication of how the exposure in an all terrain vehicle might contribute to the occurence of traumatic pathology.

"Crowding and Human Aggressiveness," *Journal of Experimental Social Psychology,* Vol. 8, November, 1972, pp. 528-548.

KEY WORDS:

Man	Setting	Relational Concepts	Outcomes
Human	*Rooms*	*Crowding, sex differences*	*Aggressiveness*

Crowding was discussed in terms of the various definitions held by both laypersons and social scientists. Crowding and population density were viewed as one and the same for purposes of the two studies reported. The relation between various aggressive behaviors of persons and levels of density in urban settings was reviewed. Pertinent prior research efforts were cited, although results conflicted as to the role of crowding on crime and various pathologies.

In this study, two experiments investigated the effect of crowding on human aggressiveness by placing groups of subjects in small or large rooms for several hours. In Experiment 1, the principal objective was to measure density changes on performance in a game that allowed subjects to behave either cooperatively or competitively. Subjects were 136 high school students (72 females and 64 males). Nine pairs of female groups and eight pairs of males were assigned to two rooms that were constructed specifically for the experiment. Half of the subjects went to a large room and half to a small room. A three-stage testing procedure was administered and a questionnaire followed these tasks. Results showed that all male groups were more competitive in small rooms.

A second experiment was created using a different measure of expressed agression, context, group of subjects, and situation. More detailed analysis of effective responses were sought. In Experiment 2, 191 subjects were divided into groups and tested within actual rooms in a university building. The subject group was quite diverse in demographic composition. Within the rooms, subjects listened to a series of five mock courtroom cases simulating actual courtroom proceedings via an audio recording. Subjects evaluated each case and selected a verdict, although for half the subjects noise manipulation was introduced. A static noise present on the tapes was intended to irritate or cause stress to subjects during the experiment. The purpose of this measure was to see what effect this noise had on the subject's verdict and on the effect of the density manipulation on the groups in each of the two rooms.

Results showed that all male groups gave more severe sentences in the small room as opposed to the large room while all female groups were more lenient in the small as opposed to large room. Mixed sex groups showed no effect of room size for either the whole group or each sex considered separately. Effective responses by the same sex groups were consistent with the same set of measures. Males were generally more positive to one another in the large room while females were more positive in the small room.

In both experiments there was no main effect of crowding although there was an interaction between sex of subject in one-sex groups and the size of the room. It was concluded that crowding does not have an overall negative effect on humans and that the effects that do exist are mediated by other situation-dependent factors. Several possible explanations for the sex-difference in crowding were presented, including: 1) the possible fact that men and women are innately different; 2) differences attributed to the fact that men require more physical space; 3) cultural screening and conditioning processes that differ between the sexes; and 4) crowding effects based on behavioral expectations that are sex differentiated. This fourth explanation was seen by the authors as the strongest possible factor in support of results concerning mixed sex groups. Future research questions should focus on the underlying factors that determine how humans respond to high density.

FREEDMAN, J.L., S. HESHKA, AND A. LEVY   83

"Population Density and Pathology: Is There a Relationship?," _Journal_
of Experimental Social Psychology, Vol. 11, July, 1975, pp. 539-552.

KEY WORDS:

Man	Setting	Relational Concepts	Outcomes
Residents	Urban	Density	No effect of density, pathology (disease)

The relationship between population density and pathology was
assessed in New York City. Experimental work with both non-humans and
humans was reviewed. The research with humans showed that density does
not have overall negative effects. The same results have surfaced for
studies covering long term effects of crowding. In nonhumans, the
emergence of pathologies in rats after a density threshold was reached
have been overgeneralized to apply to the human condition as well. The
major breakdown in this attempted link has often been attributed to the
fact that cultural screening and conditioning factors cannot be
accounted for in rats in the same contexts as those which apply to human
populations.

The study utilized 29 indices that represented important aspects of
social structure, social pathology, and population density. Data was
drawn from demographic and health statistics for each of New York City's
338 health areas. The population size of each area was similar. Of the
29 indices, two measures of population density were used: 1) the number
of persons per residential area; and 2) average number of persons per
room in all household units in the health area.

The basic results were a set of simple correlations on which all
further analyses were based. Relationships between dependent variables
and density were computed via regression equations which utilized some
subset of the independent variables to predict values of the dependent
variables. No assumptions were made as to the importance of individual
factors in sets of stepwise multiple regressions.

Although there were substantial simple correlations between density
and various pathologies, controlling for income and ethnicity caused all
relationships to disappear except for a slight correlation between density and psychiatric terminations. The findings were virtually identical for density per acre and density per room, although individual
correlations differed and the two measures were only moderately correlated overall. Within this study area density was not related to higher
juvenile delinquency, venereal disease, or hospitalized mental illness.
Also, breaking the city down into separate areas was viewed as
meaningless.

Persons' freedom to choose high density settings versus opposite
settings appeared to be a key factor in determining negative consequences of the effect. Certain high density settings are desirable while
others are not. The authors stated that at present not enough research
exists in this area from which solid conclusions may be drawn and
applied on a large scale of policy decision making. It was concluded
that density, measured as people per acre and persons per room, has
little or no independent effect on pathology.

GARDNER, M.B.   84

"Factors Affecting Individual and Group Levels in Verbal Communication,"
Journal of the Audio Engineering Society, Vol. 19, No. 7, July/August,
1971, pp. 560-569.

KEY WORDS:

Man	Setting	Relational Concepts	Outcomes
Human	*Restaurant, auditorium, foyer*	*Acoustics*	*Verbal communication*

   The levels of voice communications are affected by such factors as
distance between speaker and listeners, room configuration, acoustic
treatment, and surrounding noise level. This paper investigated the
relationship between voice levels in specific environmental settings and
the kinds of groups that are the cause of background or "babble" levels.

   The author used three kinds of physical settings to measure the
babble level created by the environment's occupants. A movie theater, a
restaurant dining room, and hotel assembly rooms were investigated and
sound levels were measured using standard auditory equipment and techniques. The theater was investigated with audiences of three different
kinds of movies and exposed to a number of different social conditions.
Measurements were made, for example, of the babble level while
background music was played and again after it had been turned off.
Similarly, the foyers were investigated during periods of low density
and high denity use and the differing levels of group noise were compared. The restaurant setting was used to measure the affects of differing acoustical treatment on the floor and wall surfaces to the
overall background noise levels.

   The study showed a direct relationship between the social settings
of the theater and foyers and the intensity of the babble levels
generated. The introduction of background music in the theater, for
example, reduced the babble level significantly. The density of foyer
use was a major factor in the increase in the background noise. The
author concluded that in the absence of extraneous influences, babble
noise increased on the order of 6 db for every doubling of a setting's
population to a maximum of 80 to 90 db. He also postulated that the
introduction of such factors as white noise, crowd density controls, and
attention arresting cues such as visual or written stimuli could be just
as effective in the reduction of babble noise as the placement of physical acoustic devices.

GASPARINI, A.   85

"Influence of the Dwelling on Family Life," *Ekistics,* Vol. 36, No.
216, November, 1973, pp. 344-348.

KEY WORDS:

Man	Setting	Relational Concepts	Outcomes
Human	*House*	*Environmental arrangement*	*Social behavior*

A number of environmental variables were investigated in order to determine if relationships exist between housing sizes and types of specific social indicators within a family structure. Specifically, the author argued that a housing environment, in combination with economic and cultural indices, will affect the satisfaction and resultant behavior of its inhabitants.

A random sample of 1,000 households were surveyed in Modena, Italy, in order to determine the demographic structure of the population and tap a measure of the samples of well-being. Well-being was measured by the degree of adult quarrelsomeness and child nervousness. It was hypothesized that crowding and perception of housing sufficiency would directly affect the degrees of well-being in the sample population. It was also theorized that the social and economic status of the various families would also map a degree of correlation of satisfaction. The sample was drawn across the spectrum of the society and included members of all economic segments and represented a variety of housing environments.

The results of the survey reinforced the supposition of the author that crowding and high perceived housing densities would adversely affect a family's behavior. A crowding index of two persons per room, for example, resulted in 66.7% of the adults quarrelling and 83.3% of the children with low school performance. It was also shown, with similar statistics, that actual housing density and perceived levels of housing sufficiency had a high correlation to family dissatisfaction. The author also related the degrees of dissatisfaction to social status and income, and, as was expected, a high correlation was shown between lower status rank and higher degree of social disruption in the housing environment.

GELLER, E.S., J.F. WITMER AND M.A. TUSO   86

"Environmental Interventions for Litter Control," *Journal of Applied Psychology,* Vol. 62, No. 3, June, 1977, pp. 344-351.

KEY WORDS:

Man	Setting	Relational Concepts	Outcomes
Consumers	*Supermarket*	*Littering*	*Non-littering behavior*

This study illustrated that recent applications of behavioral psychology can potentially contribute to the control of ecology-related behaviors. The authors reviewed relevant studies that were designed to influence a range of environmental litter behaviors.

The first experiment of this study alternated between-day and within-day manipulations of baseline and prompting conditions with subjects who entered a grocery store. Experiment 2 was designed to ascertain the influence of enviromental litter on disposal behaviors in a grocery store. Sampling bias was minimized in the present research by manipulating the amount of litter in this particular type of environmental setting.

Experiment 1 showed that specific disposal instructions at the bottom of the handbill prompted about 30% of 1,146 handbill recipients to use a particular trash receptacle. Of the 1,231 customers who received no antilitter message on their handbill only 9% disposed of their handbills in one of two available trash receptacles, and this percentage did not increase when other customers received specific disposal instructions within the same time period. The handbills contained "Specials of the Week" and were handed out at the entrances to the store by a group of undergraduate student assistants. Male and female shoppers were differentiated in this process. The three imposed conditions (disposal instructions on all, half, or more of the handbills) were alternated daily until each condition occurred once on each weekday, Monday through Friday. The number of male and female recipients in a daily sample ranged from 100 females and 100 males (on 5 days) at a low of 64 males and 70 females. The locations and containers where subjects disposed of the handbills were carefully categorized and analyzed.

In Experiment 2, half of the eight observation days occurred in the same store (as in the first experiment). On Monday through Thursday of the next week, four observation sessions occurred at another large supermarket. Minor adjustments were necessary due to the change in setting. The store was systematically littered on alternate days with 140 handbills. For the littered store condition a total of 32 out of 639 handbill recipients disposed of their handbills on the floor. However, when the store was free of handbill litter, only six of 616 customers dropped a handbill on the floor.

The authors concluded that perhaps the most economical procedure for increasing the occurrence of ecology-constructive behaviors on a large scale is to provide response-priming instructions at appropriate times, and it was demonstrated that a substantial number of individuals will comply with ecology-related messages.

GELPERIN, A.   87

"Humidification of Barracks and the Incidence of Upper Respiratory Infections," *Build International*, Vol. 7, No. 3, May/June, 1974, pp. 209-215.

KEY WORDS:

Man	Setting	Relational Concepts	Outcomes
Males	*Barracks*	*Humidity control Respiratory infection*	*Physical health*

The increased incidence of upper respiratory infection among military personnel living in a barracks environment formed the basis for this six-month study. Often, during certain months of the year, a battalion of 1000 men would hospitalize 250 trainees in excess of 750 hospital days over an eight-week training cycle due to upper respiratory infection. Consequently, a study was undertaken at Fort Leonard Wood, Missouri, in order to examine a possible linkage between humidification and the incidence of upper respiratory infections.

The subjects were military personnel living in one- and three-story barracks. The buildings were arranged in groups of ten, of which eight served as living quarters. By modifying the units to incorporate humidification and air circulation, it was possible to alter the humidity in four buildings and have the other four serve as controls in each group.

The initial phase of the study involved gathering information from the subjects living in a total of 32 barracks. The buildings were then outfitted with temperature and monitoring equipment. Four barracks were provided with humidification, each set with a different level of humidity: 40, 35, 30, and 25 percent RH. The temperature was controlled at 70° F $\pm$ 3° F in all barracks. Approximately 800 men were housed in humidified quarters during the eight training weeks, and approximately the same number of men from the same companies were housed in the non-humidified buildings.

Results revealed that humidification of army barracks decreased the incidence of upper respiratory infections, from 1.33 per trainee to 1.14 per trainee during the seven weeks of observation and collection of data. Thus the author concluded that, "it is evident that creating a reasonable physiological state in the respiratory tracts of individuals of a group helps protect the persons therein from each other."

GOODMAN, R.F. AND B.B. CLARY   88

"Community Attitudes and Action in Response to Airport Noise," *Environment and Behavior,* Vol. 8, No. 3, September, 1976, pp. 441-470.

KEY WORDS:

Man	Setting	Relational Concepts	Outcomes
Human	*Los Angeles*	*Exposure to aircraft noise*	*Political and social response*

The effects of aircraft noise were measured against a community's response in terms of subjective social attitudes and direct, political actions. The authors hypothesized that a community's response can be judged by its proximity to aircraft landing zones, but exposure to noise is only one of a number of socioeconomic factors that will determine a specific reaction.

Two hundred thirty-nine residents of areas bordering the Los Angeles International Airport were interviewed to determine their reactions to aircraft noise and catalogue their socioeconomic, residential, and political status. Specific measures of the effects due to noise were defined as ability to perform quiet activities in the home, attitudes toward the airport's responsibility to abate noise, and desire to move or relocate from the path of aircraft landings. Correlations were then performed between the various environmental measures of annoyance in order to find those measures with high levels of correspondence.

In general, it was found that 59% of the residents interviewed had been bothered by aircraft noise. A correlation between environmental measures and political activism was due to noise annoyance (.24) and

socioeconomic status (.23). It was found that 29% of the sample
affected by noise had difficulty sleeping and 55% were bothered in terms
of sedentary indoor activities. People with a higher socioeconomic status tended to be more aware of and politically active against the
aircraft noise problem than those members of the community living in
apartments or classified as transient. The authors cited a number of
past studies of resident exposure to aircraft noise and concluded that
predictors of community reaction against noise could be analyzed only in
relation to each other and not as independent measures. Their basic
conclusion was that mere proximity to a noise source will not determine
a community's reaction in isolation of such factors as home ownership,
income, social status, and length of residence in the area.

GOULD, P.   89

"Changing Mental Maps: Childhood to Adulthood," _Ekistics_, Vol. 43, No.
255, February, 1977, pp. 111-119.

KEY WORDS:

| | | Relational | |
Man	Setting	Concepts	Outcomes
Human	Sweden	Mental mapping	perception

     Mental maps were presented here as the areas of a nation that
various members of its population would prefer to inhabit. The author
presented the hypothesis that experience and age will sharpen the attitudes of regional populations about their macro-environment and that
these attitudes can be generalized into a specific regional image of
desirability.

     Eleven thousand school children and adults were asked to choose
their favorite living environments from a standardized mapping procedure
in Sweden. The subjects were asked to establish 70 different preferences for the country by ranking specific geographic sectors from 1
(desirable) to 70 (least preferred). The collective maps of the subject
were then analyzed by rank order using a 70 x 11,000 data matrix and
correlating interrespondent preferences. From this analysis isopre-
ference curves were drawn on the maps for each subpopulation of the
sample. A map, for example, emerged for each of the nine, eleven, thir-
teen, sixteen, eighteen year old and adult subgroups for each regional
area that was surveyed. A final map of preference curves was also
constructed for the entire nation by combining all answers in a common
matrix.

     The maps of each age group showed a strong preference for their own
local environment with these preferences becoming more coherent and precise as the population ages. The author found that the respondents in
the sparsely populated areas of northern Sweden had an extremely isolated and convoluted image of desirability while residents of the
southern regions of the country were quite homogeneous and matched the
national average to a high degree. Urban and governmental areas became
less and less desirable to the population as it ages and coastal zones
appeared to be the most highly valued national centers of residential
preference. The author argued that a national image of living environments could be predicted from a population and that this image would be
reinforced and localized as the population became more and more
experienced in its environmental awareness.

GRANDJEAN, E. AND W. HUNTING   90

"Ergonomics of Posture: Review of Various Problems of Standing and
Sitting Posture." _Applied Ergonomics,_ Vol. 8, No. 3, September, 1977,
pp. 135-140.

KEY WORDS:

Man	Setting	Relational Concepts	Outcomes
Workers	_Standing and sitting_	_Posture_	_Pain_

The modern work environment is composed largely of work stations
where employees either sit or stand in a relatively stable position for
the duration of the workday. The authors presented a series of studies
which match the correlation between specific seating and standing positions and the resultant incidence of muscular and intradiscal pain.

Three experiments were carried out concerning the posture behavior
of shop girls, office workers, and textile workers. Percentages of specific postures maintained during the job were recorded and questions
concerning commonly experienced pain were administered. A fourth
experiment concerned the measurement of disc pressure and electrical
activity of back muscles in a series of different seating postures
using a variety of seating configurations. The first three tests were
performed in the actual work environment, while the fourth was a laboratory test situation.

The authors found a high correlation between pain associated with
the back and neck in workers who were seated for most of the working
day. Back pain was measured in 59 percent of office workers and was
found to exist in cases where the work surface was very high in comparison to the seating level. The authors, using the results from the
fourth experiment, concluded that "an office chair should have a high
backrest, and that it should be possible to tilt the whole seat shell
between an inclination of 2 degrees to the front and 14 degrees to the
back. The tiltable seat shell should be conceived in such a way as to
be easily fixed at any desired angle by the user."

GRAVURIN, E.I. AND D. MURGATROYD   91

"Spatial Aptitude and Permutational Ability," _Journal of Psychology,_
Vol. 91, September, 1975, pp. 77-80.

KEY WORDS:

Man	Setting	Relational Concepts	Outcomes
University students	_Laboratory_	_Spatial Aptitude_	_Permutational ability_

Past studies have indicated that there exists a direct correspondence between spatial aptitude and the ability to manipulate the

sequence and arrangement of letters and words. The authors hypothesized that this correlation in abilities could be measured by both letter rearrangement and number permutation when compared to the ability to visually manipulate and juxtapose visual forms.

Two groups of students enrolled in psychology classes (N = 94 and N = 53) were used as the experimental populations. The first group was administered three standardized aptitude tests which measured spatial aptitude, rearrangement of word sequences, and manipulation of number sequences. Because of time constraints only the first two tests were given to the second set of students. Pearson correlations were used as measures of effectiveness for the experiment.

In support of the hypothesis, the results of the study showed significant correlations between spatial aptitudes and the rearrangement tasks (r = .37 for Group I and r = .36 for Group II). It was also found that the correlation between the two measures were strong among both men and women and that a direct comparison between the two permutational tasks for all subjects in Group I showed a very strong relationship (r = .67, p< .0005). "The latter result, in conjunction with the consistent relationships obtained between spatial aptitude and skill in rearranging both numbers and letters, supported the hypothesis. . . that spatial aptitude seems to be associated with an ability to permute symbols."

GROVES, D.L. AND H. KAHALAS   92

"A Framework for the Analysis of Environmental Meaning," *Man-Environment Systems,* Vol. 5, No. 2, March, 1975, pp. 95-102.

KEY WORDS:

Man	Setting	Relational Concepts	Outcomes
Recreationists	*Recreation area*	*Environmental meaning*	*Image*

A study was undertaken to identify a typology that isolated variables significant in the study of environmental meaning. A user-general population dichotomy formed the basis for analysis. The dimensions of environmental meaning were viewed by the authors as central to a person's understanding of both built and natural environments.

A simple random sample of 180 users of recreation areas in and around State College, Pennsylvania, comprised the sample group. The site for the study was Game Lands 176. This public recreation area is rapidly becoming surrounded with new development. Information concerning the general population was obtained through 1970 Census data. Of the overall population of the area, 153 were personally interviewed. The interview approach was designed to elicit three attitudinal components: 1) cognitive, 2) affective, and 3) action tendency elements. These components were directly related to a series of background (how and why) questions concerning persons' use, images, and significance attached to the recreation area. A pre-test was developed that provided a framework for the formal interview phase. A series of statistical tests measured responses obtained from the interviews. A group of judges coded responses into the three general components.

Four equations were employed in a step-wise regression: 1) the users perception (frame of reference) of Game Lands 176; 2) the users perception of public forested land in the State College area; 3) the general population's perception (reference frame) of Game Lands 176; and 4) the general population's perception of public forested land in the area. In addition to the three attitudinal independent variables employed, persons' functional perspectives and effects of certain adolescent experiences were measured. This was composed of dependent variables that mediated in person's situation-attitude interactions with the recreation area. Representative meaning variables were factors analyzed and plugged into the regression analysis.

Results suggested that recreation areas possess a variety of meanings regarding users' and non-users' environmental cognitions. Meaning was shown to be a dichotomy between tangible and intangible results of persons' experiences and the impacts on behaviors. The tangible-intangible dimension determined the value of the land to the individual. Past experience affected use patterns and subsequent meaning. The general population analysis indicated that the affective components and needs were most strongly related to the formation of meaning. Behavior patterns were directly associated with an action-use dimension and inversely associated with an emotion-symbolism dimension. It was concluded that these dimensions and typologies need to be pursued further with different populations.

GUILFORD, J.S.   93

"Prediction of Accidents in a Standardized Home Environment," _Journal of Applied Psychology,_ Vol. 57, No. 3, June, 1973, pp. 306-313.

KEY WORDS:

Man	Setting	Relational Concepts	Outcomes
Females	Controlled kitchen environment	Accident behaviors	Accident predictor variables

Predictors of accident behavior were sought that were consistently and meaningfully related to human characteristics and the attributes of a simulated home kitchen environment. The principal goal was to explore the possibility of redefining the term "accident" in an effort to develop what might be called "accident behavior criteria" as related to human factors. A kitchen laboratory was used for the study of accidents incurred by 226 female subjects who performed standardized household tasks under observation.

A simulated home kitchen environment was built in a mobile van. Every item and the tasks incurred were identical for each subject. The kitchen was 8 x 11 feet. A kitchen environment was incorporated because: 1) home accidents are responsible for more disabling injuries than any other single type of accident, 2) within the home environment, the kitchen is the most frequent site of the accident for women, and 3) it was deemed desirable to structure the experimental situation to appear as natural as possible for female subjects. Each subject individually encountered a series of kitchen related tasks. The average time required to complete these tasks was two hours. Subjects were observed via one-way mirrors.

A large number of variables were included as possible predictors of accident behaviors. Four years of driving records were obtained for a subsample of 178 subjects possessing licenses. Measures of vision were made by means of the American Optical Sight Screener, manual speed and dexterity were measured via the Employee Aptitude Survey Test No. 9 (EAS-9), intelligence was measured via the Otis Self-Administering Test, attention to detail and perceptual speed was measured via the Picture Completion subtest of the Wechsler Adult Intelligence Scale (WAIS), blood pressure was taken, use of drugs coupled with drinking and smoking habits were queried and subjects' personal health histories were explored.

Each subject received scores based on the number of: a) personal injury accidents; b) property damage accidents; c) total kitchen accidents (the sum of personal injury and property damage accidents); and d) near accidents the subjects were observed to have in the kitchen. The total accidents for each category were 714 (a); 370 (b); 344 (c); and 648 (d). Automobile records for the subsample showed that there were 36 auto accidents and 108 violations.

Significant ($p < .05$) correlations were found between automobile accidents, automobile violations, and kitchen behavioral performance criteria. A number of demographic, attitudinal, physiological, and cognitive predictors correlated significantly ($p < .05$) with both total kitchen accidents and auto accidents. Successful environmental control of exposure to hazards made it possible to extend accident criteria to include other behaviors. The results supported the "accident-proneness" hypothesis and that environmental variables lend themselves to study in controlled settings.

GUTMAN, GLORIA M. AND CAROL P. HERBERT   94

"Mortality Rates Among Relocated Extended-Care Patients," *Journal of Gerontology*, Vol. 31, No. 3, May, 1976, pp. 352-57.

KEYWORDS:

Man	Setting	Relational Concepts	Outcomes
Elderly extended-care patients	Hospital	Involuntary re-location	Adaptation; morality rates

The authors reviewed a number of studies regarding relocation of the elderly and the effects of this type of physical move upon the patients. Studies ranged in content and aim from those that focused on a person's movement from the community into a mental institution, from the community into a home for the aged, from one institution to another, from one ward to another within the same institution, or from old to new facilities. The mortality rates of patients was a central concern throughout the studies cited. Some results showed a decrease in mortality rates while other studies indicated actual increases in mortality rates after involuntary relocation.

Subjects in the present study (81 males) were extended-care patients relocated due to the planned demolition of the hospital building in which their ward was located. These subjects were followed for 21 months from the date of transfer. All patients were transferred to a newly constructed extended care unit located approximately two blocks away. Eighty four percent of the subjects were age 60 or older. A specific set of variables were monitored over an extended time period. A control group was used throughout the study. After the 21 month period the researchers obtained yearly death rates and various other data from the control population that were than compared with the same statistics for the subject group.

No increase in mortality rate was detected during the first 3-months following relocation. This is an interval usually associated with high mortality in elderly persons involuntarily relocated. In the first year after the relocation, the death rate was 33.3%, compared to an average annual death rate of 41.2% during the five years preceeding the move. At 21 months, half the relocated population was alive. These data contrast with previous studies, most of which have shown a significant increase in mortality after relocation.

There were no significant differences between survivors and nonsurvivors among the relocated group with respect to age at time of transfer, length of hospitalization prior to transfer, or mental status prior to transfer. A difference existed in the ambulatory status between groups. Of the 24 ambulatory patients, 50% died within the 12 month follow up period as compared to only 26.3% for nonambulatory patients.

It was concluded that no two studies are comparable for all factors, and that the most apparent factors in determining mortality risk relate more to the degree of environmental change and to the actual transfer procedure than to individual characteristics of the relocated populations. Marked environmental change was found to be more stressful than moderate change. It was shown that the involuntary relocation of an elderly population need not result in increased mortality.

HALL, R., A.T. PURCELL, R. THOME, AND J. METCALFE   95

"Multidimensional Scaling Analysis of Interior Designed Space," *Environment and Behavior,* Vol. 8, No. 4, December, 1976, pp. 595-610.

KEY WORDS:

Man	Setting	Relational Concepts	Outcomes
Human	*Laboratory*	*Architectural arrangement*	*Preferences*

This study was an experiment in the ways in which people are able to rate an environment and the ability of researchers to measure those

responses across multidimensional scaling configurations. "The aim of
this study was to discover whether a multidimensional scaling analysis
of the pairwise estimates of similarity made by subjects while viewing a
set of designed spaces. . . could be used to determine the criteria that
the subjects used in making their judgements. . . ."

Thirty-five adult males and females who had no formal training in
design were shown a series of color slides of 11 entrance foyers to
office buildings in Sydney, Australia. Entrance foyers were used as the
environmental stimuli because it was felt that these were relatively
familiar spaces and a good deal of design consideration concerning
lighting and space had been put into each area. The subjects were first
shown an ordinal series of each slide and then a pairwise series of two
spaces at a time. In the second series, the subjects were asked to rate
the pairs as similar or dissimilar on a 10 point scale. From this rating, a three-dimensional nonmetric variance analysis revealed a possible
rating scale of environmental features. A third rating session was then
held and, using the new three-dimensional scale as a basis for slide
comparison, the subjects were asked to make rating judgements on each of
the 11 stimuli.

The first analysis revealed that the subjects were able to make
similarity comparisons among the slides in terms of: 1) effects due to
lighting, 2) size of foyers, and 3) visual texture. From this analysis,
a rating scale of six attributes was constructed that included warm-
cool, shiny-matt, bright-dull, patterned-plain, large-small, and deep-
shallow. An analysis of the rating data obtained when the above six
attributes were rated against each slide, showed a high correspondence
between the positions of each stimulus in the original three dimensional
configuration. The eigen values for the first three dimensions in this
configuration accounted for 85% of the total variance while two dimensions were responsible for only 66%. The authors concluded that the
subjects were able to create a fairly consistent and congruent "map" of
the 11 slides and this representation was reflected in the second
responses in the judgemental task. "Judgements of interstimulus similarity analyzed by the multidimensional scaling procedures seem, therefore, to provide a valid and economical way of discovering the nature of
the subjects' view of the designed environment."

HAMILTON, P., G.R.J. HOCKEY, AND J.G. QUINN   96

"Information Selection, Arousal and Memory," *British Journal of
Psychology,* Vol. 63, Part 2, May, 1972, pp. 181-190.

KEY WORDS:

Man	Setting	Relational Concepts	Outcomes
Housewives	*Sonic environment*	*Background noise*	*Recall performance*
College		*Long-term memory*	
students		*Short-term memory*	

One major finding resulting from studies which have been conducted
regarding attention and memory has been that the information selection
strategy a person employs is a function of such factors as alterations

in the task load, the priority of the subcomponents of the task, and the level of arousal elicited. Controlling for these findings, the authors undertook a two-part experiment designed to investigate the effects of noise-induced arousal on the ability to recall a list of visually presented word pairs.

One hundred subjects, consisting of 43 British housewives and 57 students at the University of Stirling, were each exposed to four possible combinations of noise and word presentation variations. A list of ten pairs of adjectives was made up in the form of two sets of slides. On one set a stimulus word was printed in the top left-hand corner, and a response word was located in the bottom right-hand corner. The other set of slides contained only the stimulus words. Subjects were instructed that they would be requested to recall the response word at a later time when they would only be presented with the stimulus word. Each stimulus word was first presented alone for 5 seconds, and then presented with its response word for a further 3 seconds. The experiment consisted of 4 trials in which two sound levels were used and words were either shown randomly or in a fixed order. In the "quiet" trial a noise level of 55 dbc was maintained, whereas in the "noisy" trial a level of 85 dbc was maintained.

Results from the experiment indicated that the previous belief that high arousal during learning will impair recall is probably not true. Examination of the scores revealed that when order is maintained over successive anticipation trials, lists are better learned in noise. The authors hypothesized that earlier tests had erred by only presenting the list of words in a random order, and had wrongly attributed the decline in recall ability to noise, rather than to the apparently important factor of how the cues were ordered.

The second experiment was designed to examine the effects of noise in cueing long-term memory. Sixty-six undergraduates from Durham University were presented with conditions similar to Experiment I. The major difference was that they responded to one list of words, responded to another, and then responded again to the first set of stimulus words, without knowing that they were going to be retested on the original list.

Results from this phase of the investigation indicated that high arousal at the time of storage does aid in the preservation of materials in the long-term. Thus, the authors concluded that "the results from the two experiments are interpreted as demonstrating increased processing of input in states of high arousal (as produced by temporary changes in environmental stimulation)."

HANEY, W.G. AND E.S. KNOWLES   97

"Perception of Neighborhoods by City and Suburban Residents," *Human Ecology,* Vol. 6, No. 2, 1978, pp. 201-214.

KEY WORDS:

Man	Setting	Relational Concepts	Outcomes
Human	*Neighborhoods*	*Size*	*Cognitive maps perception*

The study examined the perceptions of neighborhoods among urban and suburban dwellers. It was an attempt to measure the attitudes of the dwellers in terms of environmental and social satisfaction and determine the relative scope of comprehension among each group.

Two studies were carried out in selected urban and suburban areas of Green Bay, Wisconsin. In the first study 72 households were surveyed and asked specific questions about the quality of their surroundings and the physical size of what could be described as their immediate neighborhood. The sampling was stratified along geographic, sex, and length of residence dimensions. The second survey obtained responses from 24 immediate neighbors from urban and suburban clusters. This particular survey was designed to determine the consistency among various neighbors concerning a neighborhood boundary and name as well as environmental quality.

The first study indicated that the suburban dwellers were able to conceive of an immediate neighborhood that was three to eight times the size of urban residents and were much more positive in their perception of the environmental quality. In the second study, approximately 75 percent of the respondents were in agreement on geographic boundaries of their neighborhoods but only 25 percent could agree on a specific name common to their environment. In general, the study showed that the definition and imagability of a neighborhood was present in similar degrees in both urban and suburban populations. The primary divergences between the two groups was the size of comprehension and the increased feelings of negative aspects found in the urban dweller. Neighbors agreed on immediate physical boundaries but these boundaries were not related to a common or shared name.

HARRISON, J.A. AND P. SARRE   98

"Personal Construct Theory in the Measurement of Environmental Images,"
Environment and Behavior, Vol. 7, No. 1, March, 1975, pp. 3-58.

KEY WORDS:

Man	Setting	Relational Concepts	Outcomes
Residents; *Shopkeepers*	*Urban area*	*Personal construct theory; repertory grid*	*Imagery*

This paper sought to elicit the mental images underlying the psychological environment which people form. Personal construct theory and the associated repertory grid test were suggested in a previous paper by the author as being useful both for measuring and undertaking environmental images. The present study assessed the usefulness of the repertory grid in measuring the imagery possessed by two different groups of urban residents. The previously reported paper focused on the general study of a group of female residents image of their urban environment. The data which resulted in both phases of the present study were analyzed via a modified principal components analysis in order to reveal the structure of the subjects' images. A brief review of personal construct theory and use of the grid test was included in the present paper.

The sample in the first group consisted of 20 residents of Bath, England. Lengthy interviews were conducted individually. Subjects supplied an array of information classified according to construct type. The principal component analyses yielded a number of factors that represented subjects urban imagery. The second sample group was comprised of 34 retail shopkeepers in Bristol, England. Relationships were viewed between perceptions of the "real" objective environment and persons nonobjective image of the retail environment in the two urban settings. Background characteristics of subjects were analyzed in relation to the sets of component dimensions.

The consensus imagery held by the groups of respondents was partially structured along three components. These dimensions described the physical location of elements of the image, the contribution of elements to the profitability of the shops, and the role of the shopping area where a shop was located. Results indicated that use of the repertory grid test had two major methodological advantages: 1) more sophisticated imagery maps were obtainable than had been in the past; and 2) the structure of imagery as measured by the grid was shown to vary with personal and locational characteristics of respondents. Results of both projects suggested that significant links exist between persons imagery and actual behavior in the environment. The authors stressed the need to further investigate the exact nature of these links through use of personal construct theory and the repertory grid test.

HARTMAN, C., J. HOROWITZ, AND R. HERMAN   99

"Designing with the Elderly: A User Needs Survey for Housing Low-Income Senior Citizens," *The Gerontologist,* Vol. 16, No. 4, August, 1976, pp. 303-311.

KEY WORDS:

Man	Setting	Relational Concepts	Outcomes
Elderly	*Laboratory*	*Environmental manipulation*	*Preferences*

This study was an experiment of the proposition that housing preferences can be elicited from the potential residents of low-income senior citizens complexes. It is assumed that housing needs can be determined more effectively by providing a visual medium to future residents and allowing them to make rational design choices.

The study was based upon the actual design of a senior citizens complex in the South of Market area in San Francisco. One hundred twenty three potential residents of the complex were interviewed in 17 small group sessions. The residents were presented a series of slides of existing apartment developments and asked to give their preferences. Specific environmental issues attacked during the sessions were high rise vs. low rise developments, lobby arrangement, corridor, entry doors, balconies, apartment size and shape, and common facility arrangements.

Two kinds of results were achieved in the study. First, the study provided a test of the slide technique and pointed to ways in which to improve future sessions. The authors found that keeping the number of possible alternatives to a minimum is essential to keep the session comprehensive, and that illustrations must be very explicit and direct in describing a particular concept. Sessions should not last more than 30-40 minutes. Secondly, the study provided a set of data to be used in the complex design. It was found that a sizable majority of the participants preferred modern looking structures with ground floors devoted to basic commercial services, a controlled lobby with a central mailbox area, and access to private balconies.

HASKELL, B.   100

"Association of Aircraft Noise Stress to Periodontal Disease in Aircrew Members," *Aviation Space, and Environmental Medicine,* Vol. 46, No. 8, August, 1975, pp. 1041-1043.

KEY WORDS:

Man	Setting	Relational Concepts	Outcomes
Aircrews	Aircraft	Noise	Disease

Aircraft pilots and crewmen are subjected to a multitude of environmental factors not commonly shared by the majority of individuals who remain, for the most part, earthbound. An excessive level of noise is one hazard pilots and aircrew members must contend with. This study examined the relational concepts of noise stress to periodontal disease in aircrew members.

A review of the literature revealed a multitude of effects that noise may contribute to periodontal disease. It was found that periodontal pathology and the weight of the supra-adrenals in experimental animals increased with auditory and other stress. People professionally exposed to noise are subjected to cardiovascular stress severe enough to cause pathology. Adolescents have a tendency to develop angiospasms of the peripheral vessal due to noise and vibration. There is more arterial hypertension among workers subjected to noise than among office workers or manual laborers. The reactions of bodily defense to inflammation in animals is partly inhibited with auditory stress. There is an increase in inflammatory cells in the blood with noise stress. Anxious and mentally disturbed people are more sensitive to noise, and this group as a whole is more liable to have periodontal disease.

Three groups of twenty-five men were selected for the study from the Pennsylvania National Guard. Group 1 consisted of F-102 jet fighter pilots; Group 2, pilots and crew of a four engine, propeller-driven C-121 aircraft; and Group 3, enlisted men not exposed to aircraft noise as a control. The degree of aveolar, introcepted bone loss for each subject was measured from full-mouth radiographs of all groups.

The greatest amount of bone loss (periodontal disease) occurred in crew members of propeller-driven aircraft, who were not required to wear any protective noise headgear. The jet pilots, who were required to wear protective helmets which were heavily insulated against noise, had considerably less bone loss. The data from the study suggested that there was a degree of aveolar bone loss over a period of time associated with exposure to propeller aircraft noise and vibration and negligible loss for jet aircraft noise.

HASKELL, B.   101

"Association of Aircraft Noise Stress to Periodontal Disease in Aircrew Members," *Aviation Space, and Environmental Medicine,* Vol. 46, No. 8, August, 1975, pp. 1041-1043.

KEY WORDS:

Man	Setting	Relational Concepts	Outcomes
Aircrews	Aircraft	Noise	Disease

Aircraft pilots and crewmen are subjected to a multitude of environmental factors not commonly shared by the majority of individuals who remain, for the most part, earthbound. An excessive level of noise is one hazard pilots and aircrew members must contend with. This study examined the relational concepts of noise stress to periodontal disease in aircrew members.

A review of the literature revealed a multitude of effects that noise may contribute to periodontal disease. It was found that periodontal pathology and the weight of the supra-adrenals in experimental animals increased with auditory and other stress. People professionally exposed to noise are subjected to cardiovascular stress severe enough to cause pathology. Adolescents have a tendency to develop angiospasms of the peripheral vessal due to noise and vibration. There is more arterial hypertension among workers subjected to noise than among office workers or manual laborers. The reactions of bodily defense to inflammation in animals is partly inhibited with auditory stress. There is an increase in inflammatory cells in the blood with noise stress. Anxious and mentally disturbed people are more sensitive to noise, and this group as a whole is more liable to have periodontal disease.

Three groups of twenty-five men were selected for the study from the Pennsylvania National Guard. Group 1 consisted of F-102 jet fighter pilots; Group 2, pilots and crew of a four engine, propeller-driven C-121 aircraft; and Group 3, enlisted men not exposed to aircraft noise as a control. The degree of aveolar, introcepted bone loss for each subject was measured from full-mouth radiographs of all groups.

The greatest amount of bone loss (periodontal disease) occurred in crew members of propeller-driven aircraft, who were not required to wear any protective noise headgear. The jet pilots, who were required to wear protective helmets which were heavily insulated against noise, had considerably less bone loss. The data from the study suggested that there was a degree of aveolar bone loss over a period of time associated with exposure to propeller aircraft noise and vibration and negligible loss for jet aircraft noise.

"Ergonomatic Aspects of Crew Seats in Transport Aircraft," *Aerospace Medicine,* Vol. 45, No. 2, February, 1974, pp. 196-203.

KEY WORDS:

Man	Setting	Relational Concepts	Outcomes
Pilots aircrew	*Airplanes*	*Seat design/ back pain*	*Design modification*

This paper is a discussion of the problems of aircraft seat design and the incidence of back pain in pilots and aircrewmen. Evidence suggested that the incidence of low back pain among aircrew is abnormally high and the question of seat design may be of particular significance. The author pointed out that although the progress has been slow, there have been some design advances recently and the future now looks more promising. The fact that commercial airlines such as BOAC, Air France, Swissair, SAS, and KLM, find it necessary to carry out modifications and development work at their own expense on seats already installed in their aircraft points to inadequate original design.

The paper described the anatomy and physiological factors of the human body, which are often aggravated and accentuated by the fact that pilots are often required to spend very long periods "trapped" and unable to move adequately. Some possible solutions presented in the paper were: 1) evaluation of the seat should take into account the type of flying for which it is intended (e.g. long-range, short-range, helicopter, etc.), 2) construct a basic seat which would accept add-on modules to make it compatible to two or three ethnic groups and to both sexes, and 3) pilots keeping physically active during the flight and physically fit. The author also presented numerous design considerations such as lumbar support, thigh support, seat pan contours, armrests, etc.

The paper concluded that although seat design and construction must be adequate, the pilots back health also depends on his intelligent and carefully prepared action to maintain the condition of his back such that it will withstand the inevitable natural hazards of the flying profession.

"Perceived Openness-Enclosure of Architectural Space," *Environment and Behavior,* Vol. 6, No. 1, March, 1974, pp. 37-52.

KEY WORDS:

Man	Setting	Relational Concepts	Outcomes
Students	*Laboratory*	*Boundaries*	*Space perception*

the part of the experimenter. Sixty male undergraduate students, randomly assigned to conditions, were asked to answer or not answer six
typed questions. These responses were recorded on tape. Subjects were
then asked to evaluate the experiment and the experimenter utilizing a
10-point scale of feeling. This self disclosure (time in seconds discussing questions) and feeling data were then evaluated using a 2 x 3
factorial analysis of variance.

Differences in physical and phenomonological distance had varying
effects on the subjects. Subjects talked longer in the culturally
appropriate vs. culturally inappropriate conditions. The same is true
of subjects in the close and medium vs. the greater phenomenological
distances. Though physical distances exhibited no significant effects
on feeling data, subjects at the close phenomonological distance reported more positive feelings than those at the greatest distance. No
interaction effects were evident.

HEFT, H.   105

"Background and Focal Environment Conditions of the Home and Attention
in Young Children," _Journal of Applied Social Psychology,_ Vol. 9, No.
1, January/February, 1978, pp. 47-69.

KEY WORDS:

Man	Setting	Relational Concepts	Outcomes
Children	_Home_	_Noise, archi-tectural arrange-ments_	_Attention levels_

The author of this article suggested that the amount of residential
noise and focal stimuli that young children encountered will affect
their abilities to perform certain skills and their attention spans.
Specifically, he argued that children from homes with high noise levels
would have longer response latencies and would recognize fewer incidental aspects of a particular task than those from quieter environments.

Ninety-four families of kindergarten age children were used as subjects for this experiment. The families were divided equally between
rural, suburban and urban areas surrounding Altoona, Pa. Measurements
and assessments of noise levels were performed in each setting and room
configurations and arrangements were scored with respect to variations
in colors and textures, size, and occupant density. Approximately half
the children were male and half female and variances in family socio-
economic status were taken into account throughout the analysis of the
data.

Attention experiments were performed on the subjects by administer-
ing a visual recognition test while exposed to differing levels of background noise and visual distractions. The data indicated, as theorized,
that children from noisy homes were less responsive to extraneous sounds
and less efficient in performing a task designed to assess attentional
skills. "These findings suggest that adaptation to potential environmental stressors may be only a partial process, perhaps temporarily

An experiment was performed to examine the hypothesis that the perception of openness-enclosure of architectural space is determined by the relation of boundary height to boundary distance and is independent of the actual size of space. Prior studies of space perception have been largely confined to the study of perceived size, distance, form, and motion. The authors maintained that perceptual psychologists have only recently directed their attention to some of the more complex dimensions that determine our responses to the large scale physical environment. A number of relevant prior studies were cited.

A group of twenty undergraduate university students monocularly viewed a series of 12 perspective drawings which varied in size (scale) and boundary height-distance relations. Each subject individually rated the group of drawings accordingly. The results indicated support for the hypothesis and provided empirical evidence that the experience of openness can be mediated by design and need not depend on actual extended space. Possible explanation for the results was attributed to the size-distance paradox. Increases in scale produced both over-estimation of back wall size and under-estimation of distance. Subsequently, variations in scale were partially responsible for differential changes in the value of the subjects' perceived height-distance ratio. The small but consistent effect of scale was demonstrated at all levels of ratio with the exception of the 1:1 height-distance value.

It was concluded that the data offered support for Spreiregens hypothesis regarding the impression of openness-enclosure of architectural space. Also, support was afforded to the intuitive notion that the experience of openness can be mediated by design and need not depend on the availability of actual extended space. Scale made no difference in how open or enclosed a space appeared.

HAZELWOOD, M.G. AND W.J. SHULDT   104

"Effect of Physical and Phenomenological Distance on Self-Disclosure," *Perceptual and Motor Skills,* Vol. 45, 1977, pp. 805-806.

KEY WORDS:

Man	Setting	Relational Concepts	Outcomes
Male college students	*Laboratory*	*Physical and phenomenological distance*	*Self-disclosure*

Due to the scarcity of research assessing the simultaneous effects of physical and phenomenological (i.e., self disclosure by the experimenter) distance on behavior within psychotherapeutic relationships, this study was designed to measure the effects of these dimensions within a setting analogous to psychotherapy.

The experimental design consisted of two physical distance conditions, in which chairs were placed at culturally appropriate (1.8m) and culturally inappropriate (.7m) distances, and three levels of phenomenological distance, with declining amounts of touch and self disclosure on reducing, but not completely ameliorating their long-term impact on the individual." No significant correlation was found between environmental arrangement and attention spaces as was originally proposed.

HENDEE, J.C., R.P. GALE AND W.R. CATTON, JR.   106

"A Typology of Outdoor Recreation Activity Preferences," *The Journal of Environmental Education*, Vol. 3, No. 1, Fall, 1971, pp. 28-34.

KEY WORDS:

Man	Setting	Relational Concepts	Outcomes
Human	*Parks*	*Recreation*	*Preference*

The authors of this study hypothesized that there is a correlation between a person's age and education level and the kinds of outdoor recreation that he/she will prefer. Specifically, they attempted to establish the relationships between a range of activities and the stated preferences of a group of outdoor facilities users.

Two-thousand four-hundred users of western Washington national parks were surveyed concerning their age, education level, and preferred outdoor activities that they hoped to use while in the facility. The activities were grouped into five basic categories which, based on past studies, provided a generic range of activities levels. They were Appreciative (hiking, climbing), Extractive (hunting, fishing), Passive (relaxing, canoeing), Sociable (study, visiting), and Active (swimming, skiing). These levels of activity were used to group basic types of outdoor functions around physical and psychological indices which would be useful in classifying particular population characteristics versus a preferred outdoor facility.

It was found that, in accordance with the stated hypothesis, that there was in fact, a correlation between age and education level and a particular set of outdoor recreation. It was shown, for example, that older respondents (50 or over) had increasing preferences for passive activities and decreasing preferences for active recreation types. "Preference for appreciative-symbolic activities was most typical of young adults (20-29) and least typical of those 60 years of age or older." A significant correlation was also recorded between a high education level and appreciative activities while nearly 75 percent of those preferring hunting and fishing sports had not completed college. In conclusion, the authors found that a pattern of recreation preferences could be predicted given demographic variables and outdoor facilities could be modeled around this set of information inputs.

"The Development of Cognitive Mapping of the Large-Scale Environment,"
Journal of Experimental Child Psychology, Vol. 26, No. 3,   December,
1978, pp. 389-406.

KEY WORDS:

Man	Setting	Relational Concepts	Outcomes
Children	Scale model town	Cognitive mapping of large-scale environments	Accuracy levels via repeated exposure

Two experiments were conducted to investigate the development of
children's cognitive maps of large-scale environments. The tasks presented to the subjects were designed to investigate changes in the accuracy of children's cognitive maps that occurred as a function of
specific and repeated experience. Literature was reviewed that focused
on children's: 1) development of topological spatial concepts, 2) use of
projective and Euclidean concepts such as metric distance relations, and
3) abilities to transpose cognitive imagery from actual environments to
scaled representations of what they recalled from actual environments.
The authors contended that previous results of such experiments which
indicated young children's inaccuracy and their inferred inability to
use Euclidean relations was an artifact of a fourth methodological problem: providing children with only limited experience with the environment. The few studies which were cognizant of this fourth methodological problem were cited. These findings suggested that perceptual accuracy increases as a function of the amount of sensorimotor experience
with an object or stimulus. In this study it was hypothesized that
younger children would initially have difficulty when placing landmarks
in the environment that have either ambiguous or isolated topological
positions, and that with repeated experience with the scaled layout even
the youngest children would be able to place all buildings with considerable accuracy.

Kindergarten (mean age = 5.7), second (mean age = 7.7), and fifth
(mean age = 10-7) graders walked through a large scale model town and
were then required to construct the layout of buildings in that model
from memory. Ten boys and ten girls from each of three grade levels in
a parochial school participated. In Experiment 1, a large classroom was
used to present the layout of a 4.9 ft. x 6.1 ft. area on the floor
which contained 8 scaled buildings, a road, and a railroad track. Children individually walked through the town in either of 2 experimental
conditions (Walk or Construct condition). The five children in the Walk
condition built the town from memory only after the third guided walk
through, while the five children in the Construct condition rebuilt the
town from memory after each of the three guided walks. A series of analyses of variance tests measured the levels of interaction between the
various groups, buildings, and conditions.

Accuracy of the children's reconstruction increased as a function
both of developmental level and repeated encounters with the layout. In
Experiment 1, the separate effects of repeatedly walking through the
town and constructing it immediately afterwards increased the accuracy
of the children's reconstructions.

The purpose of Experiment 2 was to assess the independent effects of walking and looking (without walking through) and the subsequent accuracy of reconstruction. In the first experiment it was not possible to determine whether repeated walking through the town contribued to the accuracy of the children's reconstruction or if this occurred because the children were able to repeatedly view the layout. Eight boys and eight girls from each of the same three grade levels from another grade school participated. The experiment was conducted in the school gym using the same town layout, although the buildings were 1.5 times larger. Four boys and four girls at each level were tested in either the Walk or Look condition or the Look Only condition. The testing and analysis procedures were approximately the same as those used in Experiment 1.

The results of Experiment 2 indicated that walking through the environment was no more effective than merely viewing it repeatedly. Young children's accuracy in bounded space (the classroom in Experiment 1) was far more accurate than their performance in unbounded space (the gym in Experiment 2) while older children's accuracy was relatively unaffected by this variable.

Cognitive maps which guided children's reconstructions improved as a function of development and experience. No sex differences were found in Experiment 1, but boys were signficantly more accuate in Experiment 2. The authors concluded that one area of future work should focus on the multidimensional scaling of children's relative distance judgements between "landmarks that they have experienced in a real-world environment."

HERMANN, E. AND H. CHANG 108

"Acoustical Study of a Rapid Transit System," *American Industrial Hygiene Journal*, Vol. 42, No. 9, October, 1974, pp. 640-653.

KEY WORDS:

Man	Setting	Relational Concepts	Outcomes
Train riders	*Trains*	*Noise/health hazard*	*Hazard determination*

This study examined noises generated by the trains of the Chicago Transit Authority (CTA) under various operating conditions. These noises were recorded in the field and examined in the laboratory. The data collected from noise analysis were further examined to evaluate the extent of occupational health hazard to crew members and those who work in the newstands located on the platforms of subway stations. Speech interferences levels and growth recovery of temporary threshold shifts (TTS) in hearing sensitivity among passengers and crew members were predicted and the probability of incurring noise-induced permanent threshold shifts (PTS) were also estimated.

The study began with preliminary investigation to select proper equipment in noise recording and analysis. Using equipment thus

selected, comprehensive noise recordings were made in the field and subsequently analyzed in the laboratory. Data obtained from noise measurements were analyzed to estimate the occupational health hazard and
speech interference among crew members and passengers.

It was found that the overall sound pressure levels of the noise
within the moving cars of CTA trains varied from 60 to 115 db. The
characteristics of these noises depend on a number of factors, including
type of car, type of trackbed construction, time of operation, window
conditions, shape of track, and train speed. Generally speaking, the
major frequency components of CTA train noises concentrated at frequencies below 1000 HZ, except for a screech generated when a train travels
on a curved track. The daily noise exposure fraction for crew members
ranged from a value far less that 1.0 to a value slightly greater than
1.0, the threshold limit specified by OSHA. Therefore, years of daily
exposure have had and will have adverse effects on a small proportion of
train crew members. Although some portion of passengers developed a
small amount of temporary threshold shift in a single trip, relatively
few developed permanent threshold shifts from this source. Speech communication among passengers was difficult in most cases and sometimes
impossible even with shouting. There appeared to be no deterioration in
hearing sensitivity of those who work in the newsstands at subway stations. Such noises, however, posed serious communication problems
important to business transactions.

HERZOG, T.R., S. KAPLAN, AND R. KAPLAN   109

"The Prediction of Preference for Familiar Urban Places," *Environment
and Behavior*, Vol. 8, No. 4, December, 1976, pp. 627-645.

KEY WORDS:

Man	Setting	Relational Concepts	Outcomes
Students	*Laboratory*	*Urban places*	*Preferences*

The authors of this study attempted to relate the preference of an
urban place to a person's familiarity and concept of complexity of that
place. They also tested the viability of using photographs of these
spaces in order to establish a congruent image in all subjects' minds.

One hundred and twenty-one introductory psychology students were
used as subjects for this test. Ten groups of the students were
instructed to rate a series of 86 scenes with which they were assumed to
be at least moderately familiar. The scenes were presented by color
slide (slide condition), as a name and location only (label condition),
and as a name and location with the instruction to image the scene as
vividly as possible for 15 seconds (imagery). Each scene was rated by
the subjects in terms of preference, familiarity, and complexity, with
the use of a 5-point verbal scale. Seventy scenes were eventually found
to be most representative of five separate categories of places labelled
generally as Cultural, Contemporary, Commercial, Entertainment, and
Campus dimensions by the use of nonmetric factor analysis of the preference ratings for the slide condition.

Significant differences in predictors of preference were found among the five visual dimensions ($F = 23.24$, $df = 4.92$, $p < .001$). The Commercial and Entertainment dimensions were the least preferred while the Contemporary and Cultural dimensions were relatively high in preference. The highest ratings of familiarity were found to be those of the Campus dimension, while the complexity index was rated highly in the Cultural dimension. It was also found that familiarity and complexity (independent of one another) accounted for 48% of the preference variance in the slide condition. The correlations of preference among the three presentation conditions were quite high (.81 to .86) which indicated the use of imagery and label conditions were equally as effective as slides as stimuli for subject mental imagery. The authors concluded that preference of urban spaces can not be predicted by form or function indices alone. The study indicated that, while Cultural dimensions were strongly linked to functional criteria and Contemporary dimensions were preferred for form characteristics, most of the dimensions suggested various degrees of mix. "Complexity emerges here as a factor in what people prefer, but not as a factor that can be viewed in isolation from other important influences. It also is not uniformly effective across all contents. Familiarity . . .was on the whole as effective a predictor as complexity, although its variation as a function of content was substantially greater."

HESS, R.F. AND N. DRASDO   110

"Seizures Induced by Flickering Light," *American Journal of Optometry and Physiological Optics*, Vol. 51, No. 8, August, 1974.

KEY WORDS:

Man	Setting	Relational Concepts	Outcomes
Epileptics	*Environment*	*Seizures/flickering light*	*Preventions and treatment*

This article reviewed past studies concerning photo-sensitive epilepsy and flickering light. The studies reviewed were conducted both in the laboratory and field. Photo-sensitive epilepsy is a pathological entity characterized by sudden recurring seizures caused by intermittent visual stimuli. Although this form of epilepsy has been recognized for a relatively long time, its etiology and pathogenesis is only gradually being pieced together. Intermittent light stimuli, occurring in the environment, results mainly from the frequency of alternating current in the main electricity supply, or the periodic physical interruption of light from a constant source.

Television is a major source of inducing seizures. The most common type of fit to occur from television is the grand mal (major seizure). The most provocative situation is when the person is at close range to the set in a dimly lit room. Also, photo-sensitive subjects travelling along a road can be affected by the flicker produced by sunlight rythmically interrupted by a row of trees. Flourescent lights in an underpass may also induce seizures in car drivers.

A photoconvulsive response involves some loss of consciousness, eye movement and specle arrest. Photoconvulsive attacks range from petit mal, myoclonic jerking and photomotor attacks to grand mal seizures. The petit mal, as the name suggests, is less severe, involving a typical first order photo convulsive response. This results in a disturbance of consciousness, specular arrest and a fixed staring appearance. The seizure duration may vary from a few seconds to a few minutes. The more severe grand mal attacks usually involve arrest of consciousness followed by tonic/clonic movement of extremities, urinary incontinence, and frothing and biting of the tongue. Therapy often takes the form of advice concerning the avoidance of potentially dangerous situations, and how to abort the attack if prodramal symptoms manifest themselves. This includes the use of conditional therapy, drugs, and various optical devices.

HIGH, T. AND E. SUNDSTROM   111

"Room Flexibility and Space Use in A Dormitory," Environment and Behavior, Vol. 9, No. 1, March, 1977, pp. 81-90.

KEY WORDS:

Man	Setting	Relational Concepts	Outcomes
Students	Dormitories	Architectural flexibility	Behavior

This study examined the relationship between architectural flexibility and resulting social behavior. The authors hypothezied that dormitory residents living in rooms with flexible furniture arrangements will have a higher index of interpersonal social behavior than those living in rooms with fixed furniture units.

Twenty male and twenty female students living in double rooms at three dormitories at the University of Tennessee were the experimental population of this study. Half the rooms were equipped with movable furniture pieces and room arrangement was left to the students' discretion while the other half were fitted with fixed bed and study built-in units and were categorized by the researchers as "non-flexible". Each subject was given a 15-minute interview and asked to rate his or her feelings about their room's flexibility and list the number and type of social situations that had occurred in the room during the last week. Unobtrusive observations on room usage were also carried out on a random basis and activities in each room were catalogued over the study period. Data were examined in 2 x 2 analyses of variances between room flexibility and occupant sex.

Occupants living in flexible arrangements rated their environments higher than those living in the non-flexible rooms ($F = 165.83$; df = 1.36; p<.001), with males having a significantly higher perception of flexibility than females. It was also found that females were more prone to have a higher incidence of interpersonal interaction in flexible rooms than males ($X^2 = 6.52$; p< .02). The authors concluded that residents in the non-flexible rooms were forced to look elsewhere for social contact and females tended to be more concerned with the arrangement of their settings during socialization. They suggested that, in general, females could be more sensitive to their physical setting than males and different space requirements may be required for male and female dormitories.

HILL, M.R. AND T.T. ROEMER   112

"Toward an Explanation of Jaywalking Behavior: A Linear Regression Approach," *Man-Environment Systems,* Vol. 7, No. 6, November, 1977, pp. 342-349.

KEY WORDS:

Man	Setting	Relational Concepts	Outcomes
Pedestrians	*Intersections*	*Jaywalking behavior*	*Social context*

Pedestrian jaywalking was the principal concern of this study. The setting was an urban, signal-regulated traffic intersection. This type of behavior was viewed holistically as environmentally situated behavior. Prior studies have generally approached jaywalking in a segmented manner, examining either the traffic engineering or the social-psychological determinants. The present study sought to assess their relative importance jointly and provide a more inclusive framework. A linear regression technique was adopted in the study.

A series of tenets concerning dimensions of urban jaywalking were outlined. Traffic law violations and pedestrian non-compliance are generally considered by society as insignificant social problems, in spite of the large numbers of road fatalities which occur annually.

Data was collected by teams of undergraduate and graduate university students through the direct observations of pedestrians and urban intersections in Lincoln, Nebraska. Students were instructed to observe the first individual who arrived at the curb during the DON'T WALK segment of the traffic signal cycle. Each student then recorded and coded data for the following variables: 1) sex; 2) age; 3) group (alone or with others); 4) offcurb (compliance with signal); and 5) day of week. The graduate student team collected additional data on traffic conditions, number of lanes, the location of the specific intersections observed, and other miscellaneous information.

Results demonstrated that linear regression analysis was useful in the explanation of the ethologically observed behavior in this environmental setting. The Chow test was utilized as a technique for the comparison of data sets collected by different observers collected at different times and places. It was concluded that a measure of confidence could be placed upon data collection by the undergraduate teams. Jaywalking appeared to be a contextually determined behavior. Knowledge of the background characteristics of pedestrians did not add significantly to an explanation of jaywalking. The role of a conceptual framwork in determining the ordering of variables into the regression equation was illustrated. The authors stressed that future efforts should focus on both the physical and social context of jaywalking behaviors in field settings.

"Behavioral and Attitudinal Effects of Large-Scale Variation in the
Physical Environment of Psychiatric Wards," *Journal of Abnormal
Psychology,* Vol. 82, No. 3, December, 1973, pp. 454-462.

KEY WORDS:

Man	Setting	Relational Concepts	Outcomes
Psychiatric patients	*Psychiatric ward*	*Environmental arrangement*	*Social behavior*

        This article suggested that significant changes in the physical
environment can affect the behavior of psychiatric patients and aid in
the treatment of socially related mental diseases. Based upon past work
concerning this hypothesis, the authors showed through experimentation
that the arrangement of space is an important factor in the ways in
which psychiatric wards are administered.

        Two identical wards of a large municipal hospital were used as
experimental settings. The control ward was a U-shaped pavilion with
open bay bedrooms, darkly painted walls, and old and deteriorated
furnishings. The experimental ward was painted with bright and highly
contrasting colors, beds were partitioned with 6 foot high screens and
new furnishings were incorporated in all public and private spaces.
Twenty-five patients were selected randomly from each ward and were statistically similar in age, sex, and mental disorder. Both observational
and interview techniques were used to measure the patients' social and
personal behavior patterns.

        The study showed that patients in the test ward were more socially
active and much less passive than those of the control ward. Significant differences between the two wards were demonstrated in the relation
between behavior and interview scores. It was found, however, that isolated active behavior did not vary between the two wards, and different
behavior patterns in the bedrooms were not found. Some observations,
however, indicated that patients tended to be more willing to take an
active part in the decoration of their own personal space in the ward
with the privacy screens. The authors concluded that the most important
finding of the study was that psychiatric wards should reflect the adaptive nature of private spaces outside the hospital and should encourage
socially active behavior.

"Environmental Effects on Outdoor Social Behavior in a Low-Income Urban
Neighborhood: A Naturalistic Investigation," *Journal of Applied Social
Psychology,* Vol. 6, No. 1, 1976, pp. 48-63.

KEY WORDS:

Man	Setting	Relational Concepts	Outcomes
Human	*Urban neighbor-hood*	*Environmental arrangement*	*Social behavior*

Experience in public housing projects has shown that a good deal of dissatisfaction has occurred with respect to the provision of adequate and appropriate outdoor space. The author suggested that the tendency of modern urban design to stress simplicity and sterility rather than complexity and diversity has resulted in the lowered use of new housing sites as opposed to the more traditional urban dwellings of the poor.

Three different types of low-income neighborhoods were examined on adjacent sites on the lower east side of New York City. The first was a typical 3-story walk-up type neighborhood that shared a public street as its common outdoor space. The second was a group of 74 story apartment towers clustered around an open, grassed block. The third environment was also a group of high-rise buildings, but which was provided with a specially designed amphitheatre, playground, and central pedestrian mall. Behavioral observations were made of 15 percent of the outdoor users of each environment over a two month period. Behavior observed was verbal and social contact, active recreation, isolated recreation, and functional exterior tasks.

In conformation with the author's hypothesis, the study showed that there was a much greater usage of outdoor spaces in the old and innovative neighborhoods for socializing as opposed to the traditional highrise environment. The amount of outdoor activity was highest in the old neighborhood, especially in the immediate area of the tenement's entrances and sidewalks, which accounted for 90 percent of its outdoor activity. Benches were the most widely used outdoor furniture, while the innovative playground was rarely used at all. Open outdoor space seemed to be the most highly sought after commodity by youths in all neighborhoods.

HOLAHAN, C.   115

"Effects of Urban Size and Heterogeneity on Judged Appropriateness of Altruistic Responses: Situational vs. Subject Variables," *Sociometry*, Vol. 40, No. 4, 1977, pp. 378-382.

KEY WORDS:

Man	Setting	Relational Concepts	Outcomes
Students	Laboratory	Urban exposure	Helping behavior

It has been suggested in previous research that there is a direct relationship between a person's hometown affiliation and his attitude toward helping a stranger in need. Specifically, it has been argued that urbanites are more aloof and unhelpful than members of rural communities. The author hypothesized in this study that attitudes of helpfulness are primarily the result of a situation rather than a person's background.

One-hundred and seventy three undergraduate students (both male and female) were chosen in relation to their size of hometown. Forty seven were from cities of 50,000 or less, 57 from cities of 50,000 to 500,000 and 69 from cities greater than 500,000 in population. Each student was asked to rate on an appropriateness scale of 0 to 9 four different

scenarios in four different geographic settings. Each scenario dealt
with the helping of an unknown 25 year old male from a high to low risk
situation. The four settings varied from a heterogeneous population of
over a million to a homogeneous population of less than 1,000. Data
were analyzed in a six-factor analysis of variance measuring appropriateness with sex, hometown, size, heterogeneity, urgency, and risk.

"Using a measurement scale of judged appropriateness. . . , results
indicated that altruism is negatively related to situational variables
such as population size, population heterogeneity, and a level of personal risk, and positively related to the urgency of the request. Among
personal variables, sex was related to altruism (higher in males), but
hometown background was not. Interactions occurred between hometown and
urban size, and between risk and a number of other variables. Findings
(were) interpreted as offering evidence that the urbanite's unrespon-
siveness is more a function of situational factors and of interactions
between situational and subject variables than of subject characteristics alone."

HUDSON, R.   116

"Images of the Retailing Environment: An Example of the Use of the
Repertory Grid Methodology," _Environment and Behavior,_ Vol. 6, No. 6,
December, 1974, pp. 470-494.

KEY WORDS:

Man	Setting	Relational Concepts	Outcomes
Consumers	Downtown shopping area	Cognition	Spatial behavior

The author of this paper assumed that a necessary condition for
studying cognitive images of environments is an appropriate method to
define and measure such perceptions and cognitions. Previously, the
mesurement problem has often proved a major stumbling block and has usu-
sally delayed the maturation of this approach to understanding environmental imagery. Studies have suggested that the Repertory Grid
methodology offers a potential solution in these measurement problems.
Using a sample of students in Bristol, England, the Repertory Grid
method was used to measure images of the retailing environment,
focusing on two main objectives: 1) the utility of the method in
measuring environmental imagery; and 2) the range and variety of cognitive dimensions used by subjects in structuring their models of the
retailing environment, the overlap in dimensions between persons, the
perceived importance of their dimensions, and the accuracy of these imagery dimensions.

In the data elicitation process, a number of sequential steps
comprise the Repertory Grid methodology: 1) the selection of grid
elements; 2) elicitation of personal constructs from individuals; and 3)
the process of subjects scaling elements along construct scales relative
to the grid elements. The diversity of discriminatory criteria employed
suggested that while time-space constraints were important in influencing choice patterns, these responses were simply one of a set of relevant influences. The student consumer emerged as a sophisticated

decision maker. Data were analyzed according to: 1) twenty-four shop attributes that comprise the set of personal constructs elicited from subjects; 2) the number and range of shops scaled; 3) the ranking of shop attributes in terms of their perceived importance in choice of preferred shop; and 4) the relationship between objective and subjective measures of selected shop criteria.

The results supported the validity of the Repertory Grid method in measuring environmental images. Also, the variety and range of dimensions used to structure the environment has important implications in terms of devising theories of spatial choice and behavior, and in terms of the policy making and planning of the environment. The author outlined possible future research directions using the methods employed in this study.

HUGHES, J. AND M. GOLDMAN   117

"Eye Contact, Facial Expression, Sex and the Violation of Personal Space," *Perceptual and Motor Skills,* Vol. 46, 1978, pp. 579-584.

KEY WORDS:

Man	Setting	Relational Concepts	Outcomes
Experimental subjects	Elevators	Eye contact, facial expression, sex differences in	Personal space violation

This investigation, which consisted of two field studies, investigated conditions in which an experimental subject violated another's personal space. Variables under investigation were facial expression, eye contact, sex of the subject and sex of the experimental confederate.

In the first experiment 79 female and 105 male subjects were forced to violate the personal space of confederates who had positioned themselves 12 inches from the two floor selection panels in the elevators of a 30 story office building. While one confederate directed his gaze at the incoming passengers, the other stood with his back to the panel and read a newspaper. Confederates were either both male or female. For male Ss increases in eye contact from male or female confederate was accompanied by a decrease in violations of personal space. When confederates were male-female confederates, female subjects preferred to violate the space of the female who had gazed at them.

It was felt that perhaps the direct gazing of the confederates was acting to extend their personal space and thereby decreasing intrusion. Experiment 2 was designed to determine whether a confederate projecting a friendly cue vs. one avoiding eye contact would experience more intrusion. The design was similar to the first experiment. However, in this case one confederate smiled at incoming passengers while the other presented his back to the panel. Subjects consisted of 90 males and 86 females. Male subjects again preferred to violate the personal space of the confederate who presented their back. Females however, regardless of the sex of the confederate, chose to violate the space of the one who smiled. It is concluded that these results in general, point to fundamental differences in the manner in which male and female personal space violations occur.

HUMPHREY, C., D.A. BRADSHAW, AND J.A. KROUT   118

"The Process of Adaptation Among Suburban Highway Neighbors," _Sociology and Social Research,_ Vol. 62, No. 2, January, 1978, pp. 246-266.

KEY WORDS:

Man	Setting	Relational Concepts	Outcomes
Upper-middle class household residents	Interstate highway, residential area	Effects of air & noise pollution	Reaction to environmental adversities

The investigators in this study used the personal interview in a random sampling of households in 1972 of an area along Interstate 495 in Washington, D.C. The study site extended one mile back on either side of the highway which was built in 1961. Most construction of housing occurred between 1955 and 1960 before the highway opened. In 1972, the average daily traffic flow by this study site was 65,400 vehicles. Residents interviewed were 188 upper-middle class heads of household or spouses, with 70% of the completed interviews from females and 79% of the interviews completed.

The investigators examined the social impact of noise and air pollution from the limited access highway. In adapting to the highway problems, neighbors tended to complain among themselves, considered moving or tried to modify their houses or yard to keep out noise and pollution. No effect was noticed on neighboring, residential mobility or use of home for entertaining.

The results suggested a sequence of reactions to environmental adversities which began with expressive behavior and sometimes ended with aggressive social action. The interviewees in this study viewed the highway as legitimate because they used it. They demanded massive earth berms or wall, however, to cut down the noise and pollution from automotive vehicles.

HUMPHREYS, M.A.   119

"Clothing and the Outdoor Microclimate in Summer," _Building and Environment,_ Vol. 12, No. 3, pp. 137-142.

KEY WORDS:

Man	Setting	Relational Concepts	Outcomes
General population	Outdoor microclimate	Thermal factors	Thermal comfort outdoor microclimate design

Thermal comfort for outdoor environments has not been the subject
of research basically because we have very little control over the
weather. Because buildings help to modify the microclimate in the vicinity, the designer can influence the outdoor environment to a certain
degree. The aim of this study was to initially explore the factor of
clothing in terms of thermal comfort, and to provide interest in the
manner with which building design can promote outdoor thermal comfort.

At three different sites in England observations of peoples' outdoor clothing in summertime were made, and observations of the microclimate were gathered. Meteorological readings for the period in question
were recorded. The effects of heat, humidity, wind, and sunshine were
correlated with 1,365 observations of the way people were dressed.
Regression analysis of the clothing type upon the air temperature was
first performed, with terms for humidity, wind, and sunshine introduced
to the equation at a later time.

Results suggested that the clothing requirement for comfort is
largely independent of age and sex. In addition, the choice of clothing
was not affected by the atmospheric humidity within the range of air
temperatures encountered. The findings regarding wind and sunshine
indicated that "the temporal sequence of the microclimate is an important factor in determining the choice of clothing. In general, there
was a strong tendency for clothing to remain unchanged in spite of
changes in the microclimate."

The authors suggested several ways in which design can help promote
outdoor thermal comfort based on these findings. In order to allow
people the choice between the degree of shelter or exposure desired,
designers could provide leisure areas with adjacent areas of sunshine
and shade. Consideration could also be given to afford protection from
the wind, but not without allowing exposure to breezes. The provision
of shady areas with exposure to breezes cannot be overlooked, according
to the authors, since when outdoor air temperature exceeds 20°C the normal person is in light clothing, and "as the temperature rises above
this value an increasing proportion will be uncomfortably warm even
though lightly clad."

HUMPHREYS, M.A.   120

"A Study of the Thermal Comfort of Primary School Children in Summer,"
Building and Environment, Vol. 12, No. 4, 1977, pp. 231-239.

KEY WORDS:

Man	Setting	Relational Concepts	Outcomes
School children	Primary school, England	Thermal factors	Methodology revision, thermal comfort

During the summers of 1971 and 1972, the thermal comfort of 641
primary school children, aged 7-9, in England was examined. The major
focus of the study was to determine if their requirements for thermal
comfort were different than those of adults or secondary school
children, since young children have a higher metabolism than adults.

Five primary schools accepted an invitation to participate in the
study. After consultation with the staff of each school, two rooms were
chosen in each, one which contained children of mixed academic ability.
Of the 641 children who took part, 262 produced records suitable for
analysis. Thermographs kept in each classroom recorded the room temperatures. After the children were in the room for at least an hour, they
were given a form to fill out regarding what clothing they were wearing,
and how they felt. Four forms were administered per day, with a total
of 40 forms being collected if the child was in school for the total
study. Teaching went on as usual during the research period, and no
restriction was placed on the use of windows or doors to provide ventilation. Care was taken before the study began to make sure that each
child understood what the form was asking.

Results were gathered on fewer than half the children in the sample
due to difficulties in using the thermal comfort rating scale reliably
enough for analysis. It appeared that for younger children in particular, a different method of investigation would be needed. Nevertheless, some insights were gained. It was found that the children were
sent to school wearing clothes warm enough for the cool mornings, and
often failed to shed clothing as the temperature rose during the day.
Incidence of discomfort, therefore, tended to be related to the temperature variation which occurred during the day, rather than to the temperature itself. Evidence indicated that temperatures of 23°- 25°C would
be comfortable at a level of activity convenient for general classwork.
There was also no appreciable difference between responses of the boys
and the girls.

HUNT, J.C.R., E.C. POULTON, AND J.C. MUMFORD   121

"The Effects of Wind on People: New Criteria Based on Wind Tunnel
Experiments," *Building and Environment,* Vol. 11, No. 1, March, 1976, pp.
15-28.

KEY WORDS:

		Relational	
Man	Setting	Concepts	Outcomes
Human	*Wind tunnel*	*Wind*	*Comfort*

This study was designed to assess whether wind speed and gustiness
are likely to be acceptable for people in various urban environments.
Previous research regarding the effects of wind upon people has mainly
concerned itself with the cooling qualities of exposure to windy environments. The effects of changes in wind conditions experienced by
those walking out the door of a tall building or around a corner has
often been overlooked by architects and planners. The authors focused
on the effects of wind on people in normal conditions rather than with
extreme wind conditions such as those that might occur on ships or on
mountains.

Approximately 40 people were involved in a three-part test which
took place in a wind tunnel at the National Physical Laboratory in Great
Britain. In the first part, each volunteer performed various tasks
while experiencing a single wind speed, with turbulence and then without
turbulence, or vice versa. In the second part, groups of volunteers

gave verbal assessments for 20 conditions of wind speed and turbulence. The final part consisted of having each volunteer walk over a plate on the floor while being exposed to a number of wind conditions. The force with which the person stepped on the plate was measured so that an indication of the forces exerted on the floor by those people walking in the wind could be obtained. Wind conditions ranged from 4 to 12.5 meters/second.

Results revealed that at a wind speed of 8.5 m./sec. a significant deflection occurs in the direction a person is walking, by about 9 cm in three steps. This deflection is great enough to possibly lead to pedestrians slipping off a pavement or losing their balance. In addition, two subjects who were over 50 years old were blown off balance, indicating that had more older people been tested, more instances of loss of balance might have been recorded. Raising the wind speed from 4 to 8.5m./sec. or increasing the gustiness at 4m./sec. had no effect. However, adding gusting conditions to a wind speed of 8.5 m./sec. produced a 25% change in the final width of the foot marks.

The authors concluded that people were more sensitive to variation in the wind than the present criteria for acceptable wind speeds would indicate. Variable wind conditions would affect people more markedly than steady winds. The authors suggested that people's reactions in a number of places with different wind conditions need to be ascertained and correlated with wind measurements around buildings so that different criteria can be established according to the type of activity going on in the area.

IZUMI, K.   122

"Perceptual Factors in the Design of Environments for the Mentally Ill," *Hospital and Community Psychiatry,* Vol. 27, No. 11, November, 1976, pp. 802-806.

KEY WORDS:

Man	Setting	Relational Concepts	Outcomes
Mentally ill	*Clinic space*	*Environmental manipulation*	*Psychological response*

Since the perceptions of space and time are distorted in specific types of mental illness, it is important that designers of physical environments are aware of the varying responses of certain patients in specific clinic settings. The author suggested that these distorted perceptions can be predicted and environments can be designed to minimize the impacts on patients in rehabilitative settings.

The study was composed of a series of psychological experiments conducted by the author as well as previous researchers in the man-environment field. Experimental results from the works of Sommer, Hall, and Gutman were collected within this article and analyzed from four perspectives. Those perspectives were self-perceptions, perception by others, space, and time perception. The basic thesis of the study was that perceptions of patients of extreme psychosis are founded on the premise that ambiguous environmental clues are heightened in negative ways and further aggravate an ability to select from various stimuli.

Specific design problems outlined in the text were related to the reation of multiple or exaggerated images in the environment. The use of parallel mirrors, for example, has been shown to compound a patient's confusion of personal and social identity. Highly reverberant spaces cause confusion in pinpointing the location of speakers in a space. The author suggested that the use of synthetic materials in the place of traditional ones (simulated wood grain plastics) caused patients to focus on the anomaly to the point of distraction. Sounds which, in a normal setting, would be considered unobtrusive (clocks and ventilation noises) were shown to be distracting when introduced into psychiatric treatment settings. He concluded the article by stating, "This situation does not mean the environments we design for patients should be bland, sterile, and stimulus-free. It does mean we should avoid esoteric and excessive illusions that many designers deliberately try to achieve. It means designing an unambiguous, coherent, and unobtrusive environment, but one that contains the contrasts and other perceptual characteristics that under normal circumstances promote feelings of stability and comfort."

JACOBSEN, C., S. HAVEH, AND T. AVI-ITZHAK   123

"Care and Cleanliness of the Environment in Public Places: A Survey of Kibbutzim in Israel," *International Journal of Environmental Studies,* Vol. 7, No. 4, 1975, pp. 263-270.

KEY WORDS:

Man	Setting	Relational Concepts	Outcomes
Israelis	*Kibbutzim*	*Social control*	*Environmental cleanliness*

The thesis of this study was that in societies with a temporary and inadequate form of social control, negative effects can be expected in the maintenance and appearance of the physical environment. The authors argued that in newly formed settlements such as the Israeli kibbutzim a high degree of environmental pollution and carelessness would be formed in the public spaces, due primarily to a lack of societal stability and authority.

Three-hundred and thirty-two kibbutzim in Israel were surveyed in 1973 in an attempt to determine the degree of cleanliness and care that was present in the environment. Active care was defined as the conscious upgrading of the environment with shrubs and flowers, while cleanliness was measured by the absence of foreign matter or refuse in public spaces. Twenty-two separate environmental spaces were identified ranging in private to public areas from a residence's front yard to a factory yard and perimeter fence. Degrees of cleanliness and care were compared to such variables as population size, seniority of establishments, and geographic location using varimax statistical techniques.

Results of the survey tended to reinforce the thesis of the research team in that the highest proportion of cleanliness and care (67 percent) occurred within the immediate area of the individual's own residence. It was also found that the seniority of the kibbutz was an important factor in the increase in public awareness of cleanliness as was the relative size of the settlement. It was found that for kibbutzim with a population of over 700, the indices of public care of spaces decreased, which led the authors to conclude that in conformity to their original hypothesis, a community which is newly formed and with little homogeneity in its composition will suffer the effects of environmental neglect.

JONES, J.W. AND A. BOGAT   124

"Air Pollution and Human Aggression," Psychological Reports, Vol. 43, No. 3, December, 1978, pp. 721-722.

KEY WORDS:

Man	Setting	Relational Concepts	Outcomes
Students	*Room*	*Air pollution, aggression*	*Increased aggressiveness in polluted conditions*

It was stressed that in recent years a need has surfaced for environmental psychologists to focus increased attention on the effects of noise and heat pollution on human aggression in environments. Little prior research has examined the effects of secondary cigarette smoke upon human aggression in environments.

Subjects were 24 male and 24 female nonsmokers enrolled in undergraduate courses. They ranged from 18 to 25 years of age. Subjects were first either angered or not angered and then were given a chance to aggress against their provoker in a Buss "aggression machine" situation in an indoor setting. While aggressing, one-half of the subjects were exposed to secondary cigarette smoke and the other half were exposed to clean ambient air.

Results indicated that, as predicted, subjects exposed to secondary cigarette smoke were reliably more aggressive than were their clean air counterparts. Also, angered subjects were significantly more aggressive than non-angered subjects. The interaction of smoke x anger was not reliable. Experimental manipulation checks proved successful since subjects exposed to ambient cigarette smoke experienced more emotional discomfort than did subjects in the clean air group. Angered subjects self-reported greater feelings of anger toward their provoker than did non-angered subjects.

The unavoidable discomfort caused by secondary exposure to the polluted air caused subjects in both environmental conditions to be more aggressive than their counterparts who were given clean air. These findings were consistent with previous studies cited by the authors. They concluded that the responsibility lies with the environmental psychologist to design preventive strategies for coping with these adverse environmental conditions.

"Residence Under an Airport Landing Pattern as a Factor in Teratism,"
Archives of Environmental Health, Vol. 33, No. 1, January/February,
1978, pp. 10-11.

KEY WORDS:

Man	Setting	Relational Concepts	Outcomes
Residents	*Airport landing area*	*Noise, pollution*	*Mental and physical health*

With the current increase in air travel a growing concern has been
the possible mental and health consequences of aircraft noise on those
people who live near airports. Studies up to this point have yielded
conflicting results concerning psychiatric difficulties and abnormal
births of residents living under airport landing patterns. This study
was designed to examine birth records for those census tracts lying
wholly or partly under the landing pattern at Los Angeles International
Airport in order to discern if further investigation of this concern
might be warranted.

All the birth records for Los Angeles County were obtained for the
years 1970, 1971, and 1972. The census tract, race, and observable
birth defects were recorded. Those census tracts which were wholly or
partly within a 90 dba contour around the landing pattern were compared
with the rest of the county.

Results showed that the rates of teratism were higher in those
tracts within the landing pattern contour. The authors emphasized that
the study did not demonstrate that jet landing noise cause teratism.
Other factors, such as metallic particles in the jet exhaust, could not
be totally dismissed. Furthermore, one potential problem regarding
studying the effects of noise was that there existed other lesser airports in the area, along with many freeways which are scattered throughout the county. The study did point to a potentially hazardous environmental consequence leading to significant public health implications.
When combined with the findings that landing noise interrupted sleep
patterns in the area, the results suggested the need for further study.
The authors recommend that there be "great caution in permitting the
elevation of noise levels in any areas until considerably more information is available."

JORGENSON, D.O.   126

"Measurement of Desire for Control of the Physical Environment,"
Psychological Reports, Vol. 42, No. 2, April, 1978, pp. 603-608.

KEY WORDS:

Man	Setting	Relational Concepts	Outcomes
Students	*Room*	*Desire for environmental control*	*Measurement scale*

This study was designed to develop a scale that measured the desire of an individual to control the physical environment. Recent efforts dealing with environmental problems have often yielded solutions which would maintain or even increase this desire for control. This issue was conceptualized as a stable psychological characteristic which varies in strength across individual construction of a measurement scale along with an assessment of its reliability and validity proceeded in two stages. In the initial stage, 21 items were selected from a 95-item inventory developed to measure value oreintations toward nature. Responses to the set of items were intercorrelated and subjected to a principal components factor analysis.

The second stage involved administration of the resultant 14-item Environmental Control Scale to additional samples of student respondents along with other measures included to assess the validity of the scale. A second sample of 79 students was given this scale along with the Environmental Response Inventory. Additional tasks were also performed by this subject group.

The resultant 14-item scale yielded a Cronbachs alpha of .83 and a test-retest correlation of .88, as well as evidence of convergent and discriminant validity. This data suggested that the desire for control over the physical environment was reliably measured. The authors concluded that whether or not this construct remains stable over long periods of time depends on the set of underlying theoretical assumptions concerning its content and aim. Future study of this topic should focus on a) the susceptibility of this value to persuasion; b) how this value may vary as a function of situation, both in the short term and the long term; and c) cross-cultural environmental differences. A series of possible experimental designs pertaining to these items was presented by the author.

KEIGHLEY, E.C.   127

"Visual Requirements and Reduced Fenestration in Office Buildings - A Study of Window Shape," *Building Science*, Vol. 8, No. 4, ·December, 1973, pp. 311-320.

KEY WORDS:

Man	Setting	Relational Concepts	Outcomes
Office workers	*Laboratory*	*Aesthetic quality fenestration*	*Visual preference*

Windows have traditionally been designed to provide adequate daylight into the depths of a room as well as to maximize pleasant views. More recently there has been a trend toward reducing the window area in order to improve the thermal stability of the building. Little research has been carried out concerning the role that window areas play with respect to psychological needs. A series of studies were undertaken in a laboratory setting to investigate the optimal area in terms of visual requirements.

A sample of thirty participants was drawn from a volunteer staff at the Building Research Station, the location of the testing. Scale models were used in which observers adjusted a variable geometry window while looking alternatively at eight different views. The subjects were instructed to imagine that they were working in an office and to adjust the window "so as to arrive at the most satisfactory arrangement." Eye levels were held constant, and comments made by the participants during and after the testing were recorded.

Based on the window settings and the comments of the observers it was determined that a view to the exterior was an important aspect of the visual requirements of a building's occupants. "View requirements appear to be best satisfied by horizontal apertures, the dimensions of which are determined primarily by the elevation of the skyline. A range of sill heights from 0.7 - 1.1 meters and window heights from 1.8 to 2.4 meters were indicated as optimum values according to the kind of view outside. Design requirements appeared to be less decisive for fully obstructed views because of the loss of important external cues, but nevertheless indicated a basically horizontal shape." The authors warned that these findings cannot be applied to all practical situations without verification, but that they formed the basis for further studies concerning visual requirements and window area.

KEIGHLEY, E.C.   128

"Visual Requirements and Reduced Fenestration in Offices - A Study of Multiple Apertures and Window Area," *Building Science,* Vol. 8, No. 4, December, 1973, pp. 321-331.

KEY WORDS:

Man	Setting	Relational Concepts	Outcomes
Office workers, English	*Controlled setting*	*Aestetic quality fenestration*	*Visual preference*

Building on the knowledge gained from a previous study, this laboratory experiment used slightly different techniques to examine visual preferences of external views. The preliminary study utilized only a single aperture and had given an indication of the optimum arrangements. The effect of different window areas had not been examined, nor was the degree of satisfaction or dissatisfaction measured. This study was designed to investigate a number of these factors.

The sample consisted of 40 staff members working at the Building Research Station in Great Britain who were chosen to represent the general population in terms of age, sex, and status. None had taken part in the earlier study. The observers were told to imagine that they were sitting in an office and were to rate thirty different window arrangements on a five point scale of satisfaction. Three different views were used: 1) a cityscape scene with a natural horizon, such as would be seen from the uppermost floors of a tall building, 2) a panorama of midground buildings, giving an elevated skyline, seen from approximately ground position, and 3) a view entirely occupied by the facade of a nearby building. These views were located behind a mechanism which enabled a series of templates to be randomly moved in and out, altering the window areas as well as the number of apertures.

Results indicated that the optimum window height is related to the type of view. A lower sill was preferred for the distant view, an intermediate level for the facade view, and a higher sill for the scene of an elevated skyline. In addition, satisfaction was affected by the area and proportion of the window, as well as the number and width of the openings. "It appears that arrangements likely to give a reasonable degree of satisfaction will consist of apertures of horizontal proportions, each of sufficient uninterrupted width to a 12-16 degree field of the view, from the center of the room and overall, occupying about 60-75 percent of the wall width." Thus, the author concluded that although many considerations such as the thermal, structural and architectural requirements must be balanced against reduced area fenestration, a range of design options which optimizes visual satisfaction has been indicated.

KAHN, M.A.Q., W.F. COELLO, AND Z.A. SALEM   129

"Lead Content of Soils Along Chicago's Eisenhower and Loop-Terminal Expressways," *Archives of Environmental Contamination and Toxicology,* Vol. 1, No. 3, 1973, pp. 209-222.

KEY WORDS:

Man	Setting	Relational Concepts	Outcomes
Man	*Expressway*	*Proximity/lead content of soil*	*Lead contamination*

This study examined two sites to determine lead concentration in urban soils: one site was a ten mile strip along the Eisenhower Expressway which contained homes, hospitals, schools, and colleges, and the other the Chicago loop-terminal expressway which forms a loop of three different expressways receiving one of the largest traffic volumes reported in the United States. Several sites from which samples were collected were chosen by the study group alongside the Eisenhower Expressway. Generally, the shoulders are raised at an angle of about 15 to 30 degrees, inclined upwards from the roadway over a distance of about 150 feet. There was grass along this 10 mile strip except in certain places which had gravel. Samples of soil (5 cm deep) were collected at distances of 10 to 50 feet from the line marking the inner boundary of the 13 foot wide emergency lane of the road. These were taken from areas near exits, merging lanes, and areas between them.

Along the Chicago loop-terminal expressway system, most of the shoulders are made of concrete except in a few places which either have wood fillings or grass and bushes. Pigeon litter, organic matter, and vegetation is much more concentrated on the shoulder in the loop than along the Eisenhower expressway.

The soil sample taken from both expressways were dried in an oven at 150 degrees for 2 to 3 hours. The samples were analyzed for lead

either as taken or after ashing at 600 degrees for 4 hours in an
electric furnace.

The results of the study indicate that the levels vary as a function of the average distance of the soil sampling areas from the road,
and with seasons during which the soils were sampled.

Soil samples collected during March (Spring) from the adjacent city
streets showed a range of 520 to 2,230 ppm (parts per million) lead and
those collected during December and January (Winter) had lead content of
10 to 510 ppm. The lead contents of soil samples collected during
August through October (Fall) showed a range of 110 to 1,000 ppm; this
range of value is larger than the Winter range and is smaller than the
Spring range of values.

The lead levels of the soil is highest near the expressway, being
as much as 2,530 ppm, and decreases gradually with increase in the
distance from the expressway (420 ppm at 135 and 150 feet from the
expressway).

The following factors can also affect the lead levels in soils: 1)
differences in soil types and texture; 2) presence of litter, filling
materials, and living and dead vegetation; 3) traffic conditions, e.g.,
volumes and type of traffic per unit time and special conditions such as
regions near the merge lanes where acceleration takes place and those
near the exit lanes when slowing occurs; and 4) weather conditions,
e.g., wind speed and direction, humidity, rain, and snow, which affect
driving speed, the fate of exhaust fumes, and traffic volumes. The
authors noted, however, that at any one place, one or more of the above
factors can be more important than other factors in regard to the contamination of soils with lead.

The authors suggest that public buildings, houses, etc., should be
built at safe distances from the expressways and much wider buffer zones
should be provided if the slow poisoning by lead of human beings and
animals is to be stopped.

KIMBER, C.T.   130

"Spatial Patterning in the Dooryard Gardens of Puerto Rico," *Geographical Review,* Vol. 63, No. 1, January, 1973, pp. 6-26.

KEY WORDS:

Man	Setting	Relational Concepts	Outcomes
Puetro Ricans	*Entry gardens*	*Terrioriality*	*Social stratifica-tion*

The front door of a house is a clue to the social and economic status of the owner. In certain cultures, the front door is represented by
the arrangement of the floral landscape that surrounds a house. This
article categorized a number of different dooryard gardens and measured
their effects on human privacy and social status.

Eighty houses were surveyed in Puetro Rico using aerial photography and on-site observation techniques. The author recorded different types of plant arrangements and physical configurations surrounding the houses and made observations of the kinds of social activities that occurred in each type of landscape arrangement. A total of six distinct types of garden configurations were noted, ranging from the informal and natural peasant hut to the highly formal and planned plantation manor. Each succeeding category represented an increased degree of human control over the environment. Twenty-five activities were listed and observation records were kept for each of the 80 houses.

The author found that the houses could be classified as either vernacular or high-style by observing the activities of each garden arrangement. Houses with little or no human design tended to have little differentiation between public or private space, the gardens being an extension of the interior household activities. At the other extreme, houses with highly planned garden spaces tended to have a high degree of social terrioriality and a well defined sense of indoor and outdoor activities. These classifications of house types were a direct match of the economic status of the occupants. The author concluded that the degree to which a household garden encourages or discourages social communication between the occupant and a visitor provides a clue to codified cultural patterns and economic stratification.

KLASS, D.B., G.A. GROWE, AND M. STRIZICH   131

"Ward Treatment Milieu and Posthospital Functioning," *Archives of General Psychiatry*, Vol. 34, No. 9, September, 1977, pp. 1047-1052.

KEY WORDS:

Man	Setting	Relational Concepts	Outcomes
Psychiatric patients	*Psychiatric hospital*	*Social structure, hostile impluse control, time-space structuring*	*Adjustment, recidivism rates*

Studies have indicated that the ward treatment environment plays a role in the behavior of individuals while they are in the hospital. It is generally assumed that this environment also affects the behavior of the individual upon his return to the community. However, confounding variables such as the pretreatment characteristics of the patients, the length of stay in the hospital environment, and the characteristics of the community setting in which the patient was discharged have made proof of this assumption difficult. The authors examined this problem by utilizing a methodology designed to reduce the influence of some of these variables.

Two treatment wards within the same mental health center provided the setting for the study. The staff-patient ratio, treatment programs, and the use of medication on the two wards were similar. Patients who were thought to require shorter hospitalization were assigned to unit one (consisting of 7 wards), while those whose illness was believed to be more "chronic" were assigned to unit two (consisting of 14 wards).

The sample of those discharged between 1971 and 1973 consisted of 431 patients ranging in age from 19 to 60 years. Of the patients, 89.7% were labelled "schizophrenic", and 98% of all patients were released on major tranquilizers.

"In contrasting units one and two, discharged patients from unit one were found to be significantly younger (37.6 vs. 40.0 years), have a lower age at first admission (28.5 vs. 30.8 years), have more previous admissions to the hospital (4.9 vs. 3.8), and have spent less time in the hospital (10.0 vs. 12.5 weeks)." In terms of duration of time out, however, no significant difference was found between the two units. In addition, the wards within unit one did not differ in comparison to each other when measured by duration of time out. The 14 wards of unit two did, however, differ significantly from one another. "Wards that kept patients out of the hospital longer differed on the Ward Atmosphere scale from the less successsful wards in a dimension of social structure related to hostile impulse control and time-space structuring." The authors concluded that "if 'disorganized' patients arrive on highly organized wards and internalize some of the social structure of the ward, then they may stay in the community longer after discharge."

KNAVE, B., H.E. PERSSON, M. GOLDBERG, AND P. WESTERHOLM   132

"Long Term Exposure to Jet Fuel: An Investigation on Occupationally Exposed Workers with Special Reference to the Nervous System," _Scandanavian Journal of Work Environment and Health,_ Vol. 3, No. 3, September, 1976, pp. 152-164.

KEY WORDS:

Man	Setting	Relational Concepts	Outcomes
Workers	Factory	Air pollution, health	Disease

This study was concerned with the neurological and neurophysiological health examination of 29 aircraft factory workers chronically exposed to jet fuel vapors. At the factory from which the subjects of this particular study worked, the personnel were exposed to jet fuels during the production and installation of fuel systems of planes in isolated test cells, in a specially constructed fuel rig and during work on the fuel system in the complete airplanes (adjustments, repairs, etc.) at several workshops within the factory.

The authors used employee records and interviews of the personnels' own description of their earlier and present work conditions. Thirty-two employees exposed to fuel fumes from 1955 were selected for investigation (three were subsequently eliminated because of other illnesses which might bias the results of the study). The exposed persons were divided into two groups with regards to degree of exposure, one being heavily exposed and the other less heavily exposed. These classifications took into consideration both the duration and intensity of the exposure. The methods of gathering data for this study included a personal history evaluation, neurological examination, conduction velocities in peripheral motor nerves, and sensation thresholds of vibration in the extremities.

All 13 persons examined in the heavily exposed group and seven of the 16 in the less heavily exposed group stated that they had repeatedly experienced acute effects (dizziness, respiratory tract symptoms, heart palpitations, a feeling of pressure on the chest, nausea, headache) of the jet fuel vapors in the inhaled air. A high rate of symptoms indicative of neurasthemis and polyneuropathy was observed both in the heavily exposed group and the less heavily exposed group. The authors concluded that there did appear to be a correlation between long-term exposure to jet fuel vapors and pathological changes to the nervous system.

KOEPPEL, J.C. and K.W. JACOBS   133

"Psychological Correlates of the Mobility Decision," *Bulletin of Psychonomic Society,* Vol. 3, 1974, pp. 330-333.

KEY WORDS:

Man	Setting	Relational Concepts	Outcomes
College students	*Laboratory*	*Past mobility, sensation seeking, extraversion*	*Mobility*

Though there apparently have been no published psychological investigations of the personality variables which relate to geographic mobility, it is reasonable to assume that such relationships may exist. In conceptualizing this research planned mobility was used as an indication of the mobility decision and mobility was seen to exist as a continuous dimension (an individual planning to move to another state is seen as being more mobile than a person planning to move to a nearby city). With this conceptualization in mind, two studies were carried out. The first related a number of personality and biographical variables to the mobility decision, the second was an attempt to determine if results were replicable in a different sample.

A total of 160 non paid student volunteers were used in the studies. There was no overlap in subjects and their mean age was about 20 years.

Subjects in the first sample (N= 60) were administered a sensation seeking scale (SSS), the Eysench Personality Inventory (EPI) and a biographical data blank. This data blank in addition to his graphical data asked subjects to place themselves in one of four groups in relation to future mobility plans. Subjects in the second group were not administered the EPI and were administered a shortened version of the biographical data blank.

The data were analyzed using interactive correlation and regression techniques. The use of all eight predictor variables accounted for 48% of the variance. Predictor variables and multiple R squared are as follows: 1) years lived in state $R^2= 0.2119$, 2) EPI (extroversion) $R^2= 0.3217$, 3) number of organizations belonged to $R^2= 0.3631$, 4) sensation seeking score $R^2= 0.4044$, 5) family size $R^2= 0.4355$, 6) grade point average $R^2= 0.4629$, 7) EPI (neuroticism) $R^2= 0.4750$, 8) number of college attended $R^2= 0.4826$. Regression weights obtained in the model were applied to the raw scores of 10 subjects in a hold out group to provide cross validation. The correlation between actual and predicted mobility scores for thes subjects was $R^2= .63$.

It appears that the potentially mobile person in this study has a history of being mobile, is somewhat introverted, has a high optimal level of stimulation and possibly higher grades than non mobil individuals. The results of these studies strongly suggest that there are both biographical and psychological correlates to geographic mobility.

KONEYA, M.   134

"Location and Interaction in Row-and-Column Seating Arrangements," _Environment and Behavior_, Vol. 8, No. 2, June, 1976, pp. 265-282.

KEY WORDS:

Man	Setting	Relational Concepts	Outcomes
Students	_Classroom_	_Seating arrangement_	_Social interaction (verbal behavior)_

The purpose of this study was to establish the relationship between seating location in a classroom setting and verbal behavior. The author sought to prove that 1) individuals with high verbal behaviors will select a centrally located study area and 2) among individuals designated as "equal" verbalizers, those seated in central locations will exhibit a higher degree of social interaction.

One-hundred and thirty-eight students in seven separate communications classes were used as the sample population for this experiment. In each class, a measure of each student's verbal behavior was measured in what was considered a neutral (circular) seating arrangement to determine their a priori verbalization rates. Each subject was then asked to select a preferred seating arrangment in a 5-row by 5-column arrangement in order to measure their desire to seek a central physical position. Finally, each subject was assigned randomly to the 5 x 5 seating arrangement and measurements of verbal activity were collected in order to determine the effects of location on actual verbal behavior.

Results indicated that high verbalizers tended to select central seating locations (71%) at a much higher rate than did either low verbalizers (44%) or moderate verbalizers (65%). At the .05 level of confidence, this divergence indicated a significant degree of variance which tended to support the first hypothesis. In the second experiment, it was found that people with moderate verbal behavior showed a significantly higher degree of social interaction when seated in a central location. This conclusion was also drawn at the .05 level for high verbalizers, but was not shown significant for those with low verbal behavior levels. In conclusion, the author indicated that there existed significant relationships between seating location and resulting verbal behavior and that, in fact, a "triangle of centrality" could be constructed in the 5-row by 5-column arrangement which contained the highest proportion of both the high verbalizers as well as the greatest number of verbal contributions.

"Response to Altruistic Opportunities in Urban and Nonurban Settings,"
The Journal of Social Psychology, Vol. 95, April, 1975, pp. 183, 184.

KEY WORDS:

Man	Setting	Relational Concepts	Outcomes
Human	Urban	Environment	Altruistic response

    Studies have suggested that people who live in large and highly
urbanized environments are more likely to be less helpful and civil to
strangers in distress than in nonurban areas. The authors of this study
hypothesized that there was a direct relationship between city size and
the degree of incivility and cooperativeness in three specific field
situations.

    The three experiments were performed to test the above hypothesis
given an urban or nonurban setting. Forty randomly selected subjects
were phoned under the pretext that a wrong number had been dialed and
the caller needed their assistance in locating the correct number. In
the second experiment, 40 store clerks were given excessive change for a
small purchase and were given the opportunity to correct the error
before the patron left the store. A final study was conducted by dis-
tributing 36 stamped postcards and measuring the rate that strangers
picked them up and mailed them without prompts. The three experiments
were performed in urban Boston as well as in nonurban areas of Cape Cod
and western Massachusetts.

    In all three experiments, there were significantly higher likeli-
hoods of helpful behavior in the nonurban areas when compared to the
urban setting ($x^2 = 5.5$, $p < .02$). The authors concluded that the "urban
incivility" hypothesis was generally upheld and that the urban environmental factor had a demonstrable effect on the response a person was
likely to give a stranger in need.

KOSHAL, R.K. AND M. KOSHAL   136

"Air Pollution and the Respiratory Disease Mortality in the United
States: A Quantitative Study," *Social Indicators Research,* Vol. 1, No.
3, December, 1974, pp. 263-278.

KEY WORDS:

Man	Setting	Relational Concepts	Outcomes
Human	United States	Air pollution	Disease (health)

    Past studies have shown that air pollution has a negative effect on
the health of urban dwellers. The authors of this study presented a
quantitative model that attempted to specify a range of social and economic consequences of various levels of air pollution.

Data on air pollution indices and health statistics were drawn from records of forty U.S. cities during the years 1960 to 1967. The authors established a mathematical model relating the effects of suspended particulates and benzene, population density, sunshine, and humidity to respiratory mortality. The model was calibrated by applying multiple regression analyses to each index of pollution and measuring their results to actual health statistics. By checking the output of the model over an eight year cycle, the researchers were able to adjust the pollution indices to measure the effects of lowering or raising the hypothetical pollution levels on a uniform health standard.

The results of the mathematical simulation revealed that a 100 percent increase in the air pollution level would result in a 51 to 58 percent increase in respiratory mortality rates. It was also found that a similar increase in population density would increase the death rate by 12 percent while 100 percent increases in sunshine and humidity would decrease mortality in the first instance 51 percent and increase it in the second by 56 percent. The authors noted that the sunshine and humidity indices, while independent variables in the model, would be, to a large extent, tied to the effects of air quality. Taking 1967 expenditures
on respiratory disease, the researchers concluded that reduction of 50 percent in air pollution would have resulted in a social savings in the neighborhood of $1.9 to $2.2 billion annually.

KRAUSS, R.M., J.L. FREEDMAN, AND M. WHITCUP   137

"Field and Laboratory Studies in Littering," *Journal of Experimental Social Psychology,* Vol. 14, January, 1978, pp. 108-122.

KEY WORDS:

Man	Setting	Relational Concepts	Outcomes
Students, pedestrians	Urban setting, laboratory	Situational variables	Littering behavior

Four studies of littering were reported. Three different types of normative rules that govern specific social behaviors were outlined: 1) rules whose violation has immediate serious consequences and whose precise form is non-arbitrary; 2) rules whose specific form is arbitrary or conventional, yet have equally serious consequences for violators; and 3) rules that are non-arbitrary in form, and the consequences of their violation are not particularly serious. Littering was viewed by the authors as falling within this third category of normative rules. Littering and noisemaking are commonly seen as minor antisocial behaviors. Other littering behaviors were studied via observation techniques in both environmental and controlled settings. Theoretical notions relevant to the processes of compliance with and violation of minor norms were discussed.

In Experiment 1, 1,765 subjects were observed in nine different locations in New York City. This study was designed to ascertain some of the demographic and situational correlates of littering and to determine the relationship between the size of an unwanted leaflet and the probability of the object being improperly discarded. Results indicated that littering rates varied substantially across different areas of New York City and that the rate for a particular area correlated with the amount of litter already present in a particular area. It was also found that males littered more than females and young people more than old.

Experiment 2 examined, under controlled conditions, the effect of the cleanliness of an area on the likelihood that it would be littered. Subjects were presented with an opportunity to litter in an area that was either littered or clean from the outset. Thirty undergraduate students participated as subjects. The recorded behavioral measure was whether a subject littered or not in either the littered or non-littered spaces. A causal relationship between the amount of litter in an area and the likelihood that it would be littered was demonstrated. This finding supported the earlier field observation that an individual is less likely to litter a clean street than a dirty one.

Experiment 3 focused on environmental stress in the same setting as in the previous experiment. In addition, subjects were exposed to two noise conditions. Half of the subjects were exposed to loud, aperiodic noise during a task, while the other subjects were not. Thirty-five undergraduate subjects participated in this experiment. For approximately half the subjects the testing space was littered and for half it was clean. This replicated relevant findings from Experiment 2, with the exception that a relationship was not found between the amount of stress experienced by a subject and the chances that the person would litter. Thirty-two percent of the subjects littered in the pre-littered environment and 6% in the clean environment.

Experiment 4 sought to influence the level of littering in two urban field settings. The procedure was similar to that used in Experiment 1 except for manipulations necessary to increase the salience of social norms. Subjects received a leaflet and were observed as to how they disposed of it. Three petitions of varying content (all concerning the need to keep streets clean) were presented to subjects (one type per subject). The contents of each petition varied in relevance to antilittering social norms. Also, a control condition was employed. The rates for control subjects in Experiment 4 were similar to the rates found at the same locations a year earlier (Experiment 1). It was found that subjects who were approached and asked to sign the petition littered less than control subjects.

It was concluded that the most effective approach towards solving the littering problem in urban areas might be to promote positive anti-littering social norms in the classroom and via children's television programming since federal and city antilittering programs have generally been ineffective. Also, situational variations appeared to have a profound impact on littering behaviors. Other types of normative behaviors besides littering are also influenced by anti-social pressures. Person's actions in such settings were seen by the authors as contributing to the overall differences between an orderly vs. disorderly society.

"Controlled Evaluation of a Hospital-Originated Community Transitional
System," _Archives of General Psychiatry_, Vol. 34, No. 11, November,
1977, pp. 1331-1340.

KEY WORDS:

Man	Setting	Relational Concepts	Outcomes
Psychiatric patients	_Psychiatric hospital, community housing_	_Patient self-management, group therapy, planning_	_Adjustment_

Alternative treatment programs to traditional institutionalization
for mentally ill patients have grown in popularity during recent years.
Programs which emphasize partial hospitalization, community housing, and
quicker discharge rates to the community itself are becoming more common. This study was designed to compare the effects of a treatment program which provided inpatient care, a day hospital, community housing,
and sheltered work to a program emphasizing a rapid discharge policy.

Two groups of adult mentally ill patients were randomly assigned to
one of two treatment wards at the same hospital in Palo Alto, California. The experimental group (E) lived in an environment based on the
theoretical goals of designing work and living arrangements which
matched each individual's abilities, and which allowed a great deal of
patient participation in the planning and design of those living
arrangements. Consequently, group therapy and patient planning groups
were abundant. Three large houses in the community, each with its own
"atmosphere" or "lifestyle," resulted in the patients being required to
manage everyday tasks and problems. The control ward (C), on the other
hand, was based on the practice of rapid discharge and extensive pre-
discharge planning. "It relied heavily on a ward team approach, careful
use of medication, and individual consultation with the psychiatrist,
psychologist, and social worker." Work activities were also emphasized
in this program, but C patients did not enter into decision making as
did the E patients.

After discharge, 44 patients were randomly selected from the E
program and 50 from the C program. Questionnaires were administered to
the subjects after nine months and again after eighteen months. Outcome
data collected after eighteen months revealed small but significant differences between the two groups in terms of employment, maintenance of
treatment contact, use of medication, and social adjustment. Although
program effects varied considerably according to patient type, the
number of patients from the C group who returned to the hospital after
the fourteenth month was greater than for the E group. When the cost of
treatment per randomized patient was computed it showed $4,359 for C and
$6,096 for E. The authors concluded that "since both cost and effec-
tiveness were greater for E than C, the selection of the 'better' program involves value judgments beyond the scope of this paper."

"Mental Health and Population Density," *Journal of Psychology,* Vol. 85,
November, 1973, pp. 171-177.

KEY WORDS:

Man	Setting	Relational Concepts	Outcomes
Human	*Urban areas*	*Population density*	*Mental health*

    The exposure to distinct and varying environmental settings may be
a direct cause of psychotic or socially deviate behavior. The author
hypothesized in this study that an exposure to a greater degree of population density caused a proportionally increasing risk to mental health
in terms of drug addiction (social deviancy) and psychosis.

    Two demographic studies were performed using gross data and records
from a number of specific surveys which were conducted in the mid-1960s.
In the first, admittance figures to mental health institutions and private practitioners were compiled for an area of rural Minnesota and compared to similar data collected in New York City. Data was also
analyzed from studies that examined the admittance rates over a two year
period to the New York State mental health hospitals. The second study
compared the incidence of drug violations across the United States and
compared these data to corresponding population densities. Data was
collected for a one year period in the early 1960's and included all
persons charged with nonfederal narcotic law violations.

    In both surveys, the author found a direct correlation between both
mental illness and social deviancy and increasing degrees of population
density. The first study showed a ten-fold increase in mental health
admittance between the rural and urban populations (one percent vs. ten
percent) and state-wide figures of public health admittances showed a
monotonically increasing function between mental illness and population
size. The second survey reinforced the notion of social deviancy being
caused by exposure to highly congested environments in that drug arrests
varied from 2.8 per 100,000 in areas of less than 10,000 inhabitants to
15 for 100,000 to 250,000 to 61.7 for cities in excess of 250,000
people. The author concluded that both forms of social pathology are
consequences of fear and anger reactions and persons exposed to a higher
degree of negative stimuli (i.e., noise, crowds, congestion, are more
likely to withdraw into these two forms of escape.

"Crowding and Cognitive Control," *Journal of Personality and Social
Psychology,* Vol. 38, No. 3, 1977, pp. 175-182.

KEY WORDS:

Man	Setting	Relational Concepts	Outcomes
Shoppers	*Supermarkets*	*Crowding*	*Performance*

The effect of crowding on task performance is hypothesized to reflect an improved performance on simple tasks and less successful performance on complex ones. Two components of such experiences are attentional overload and behavioral constraints. When large numbers of people occupy a rather restricted space while engaging in individual tasks, the amount and complexity of situationally relevant information can lead to attentional overload. The emotional and cognitive effects of such overload seem to be more serious when the subject is engaged in tasks requiring an understanding of, and movement through the environment. The overload perspective emphasizes the cognitive involvement of the subject in experiences of crowding.

It is thought that when behavioral control is limited, individuals may still reduce the experienced adversiveness of a stressful situation through cognitive control. This may be accomplished 1) through a belief in control even when instrumental responding is unavailable; 2) through cognitive reappraisal of a threatening event and; 3) by having information about physiological symptoms that allows for the exploration and validation of experience.

Subjects for this study were recruited outside a supermarket at crowded and uncrowded times. They were given long shopping lists and told to choose the most economical product for each item. They were allowed 30 minutes to make as many selections as possible, then measures of their reactions to the situation were taken. Half of the crowded and half of the uncrowded subjects were given information that many people feel somewhat anxious and aroused when supermarkets become crowded. It was presumed that this information would increase task performance and this would be especially true in crowded conditions.

The behavioral measures included the number of items that the subjects completed and the number of incorrect (uneconomical) choices made. The post-experimental questionnaire given to the subjects was made up of questions concerning 1) satisfaction with the grocery store; 2) difficulty in finding products; 3) feelings of comfort in the store; 4) other customers getting in the subject's way; 5) difficulty in deciding on products; 6) crowdedness of the store; and 7) information about habitual shopping patterns.

Findings showed that crowding interferred with task performance. Significantly fewer items were listed in the crowded than in the uncrowded condition. Subject's shopping efficiency was also affected by crowding. The measure of efficiency used was the number of items correctly completed. The mean number of correct items for the crowded condition was 8.43, as compared to 14.55 for the uncrowded conditions. Subjects in the crowded conditions also responded more negatively to all items on the questionnaire. They felt less satisfied with the supermarket. They rated the items as more difficult to find. Crowded subjects also were less comfortable and felt that others got in their way more.

The findings of this study strongly indicated that high-density conditions may be adversive and that the adversiveness may be ameliorated by certain kinds of information. The information about possible reactions to an environment may not only make a person feel better, but may actually increase the attention available for tasks.

"Air Pollution, Climate, and Home Heating: Their Effects on U.S.
Mortality," _American Journal of Public Health,_ Vol. 62, No. 7, July,
1972, pp. 909-917.

KEY WORDS:

Man	Setting	Relational Concepts	Outcomes
Man	_Urban environment_	_Pollution/health_	_Pollution abatement_

This study examined if changes in air pollution have an effect on
health. Climate and home heating variables were included to see if they
may be pertinent. The basic approach of the study was the explanation
of differences in the mortality rates among U.S. cities by comparing the
respective levels of air pollution and socio-economic variables. The
aim of this work was to estimate the benefits of pollution abatement
programs.

The authors assumed that air pollution had important effects on
mortality, even when socioeconomic variables were controlled. In their
present analysis they added sets of "heating" variables in order to
investigate the importance of the indoor environments on mortality and
to examine the interactions of the heating variables with the pollution
variables. They also added climatic variables to examine their influence on the observed relationships.

A number of regression analyses were performed on different variables. In general neither climate nor home heating variables caused the
air pollution variable to lose significance. While there were individual pollution coefficients which did lose significance, the coefficients
were quite stable. An exception occurred when home heating fuels were
added. These variables were associated closely with measured air pollution, and both pollution and heating fuel variables tended to become
insignificant. Apparently the type of fuel used for home heating was a
major contributor to the air pollution level in the city. The socio-
economic variables were correlated with climate and home heating variables. Climate and home heating variables interacted very little in the
regressions, although there was some indication that the variables may
act as surrogates for each other. For most of the mortality rates, the
set of heating equipment variables added significantly to the explanatory power of the regression. Generally, the types of equipment were
associated with decreased mortality rates. The air conditioning variable was seldom important.

The objective of this study was to determine whether climate and
home heating would cause the estimated effects of air pollution to fall
and become statistically insignificant. In general, the air pollution
variables were quite stable. The authors suggested that there is a
close association between mortality rates and air pollution. This work
strengthened other conclusions from different studies that mortality
rates could be substantially lowered by abating air pollution.

"Residential Mobility and Creative Thinking Amoug Eighth-Grade
Students," *Journal of Genetic Psychology,* Vol. 121, December, 1972, pp.
325-327.

KEY WORDS:

Man	Setting	Relational Concepts	Outcomes
Students	*Classroom*	*Residential mobility*	*Creativity*

       The aim of this investigation was to investigate whether the fin-
dings from a previous study (MacKinnon, 1962) could be generalized to
another population (a junior high school group). The prior study
suggested that a relationship existed between frequency of moving and
the development of creativity. It was noted that the highly creative
architects studied had often moved as children "within a community, from
community to community, and even from country to country." MacKinnon's
subjects had been uprooted and forced to make their way as outsiders,
quite different from the secure childhood often thought of as optimal.
He concluded that these architects had developed personal "ideational,
imaginal, and symbolic processes."

       Eighth grade students (N = 293) were used in the present study.
Individuals were asked to record all places where they had ever lived.
The total sample was subdivided into 4 groups. The Torrance Test of
Creative Intelligence was administered. Relationships between mobility,
sex and four creativity scores were analyzed. A series of analytical
tests yielded non-significant relationships with the exception of the
differences between sexes (F = 4.03, df = 4/55, p<.01). The verbal
originality score showed the greatest sex differences in that girls
scored much higher than boys in this portion of the overall test
battery.

       It was concluded that MacKinnon's findings could not be generalized
to the group of eighth grade students. Possible reasons for the lack of
relationship were that: 1) the prior study used professional productivity as the major criterion for creativity, whereas the Torrance Tests
might be poorly or totally unrelated to creativity in architectural
design and practice; and 2) relationships that apply to a highly select
professional group such as architects do not necessarily hold for
another population.

"The Generality of Housing Impact on the Well-Being of Older People,"
Journal of Gerontology, Vol. 29, No. 2, March, 1974, pp. 194-204.

KEY WORDS:

Man	Setting	Relational Concepts	Outcomes
Elderly	*Housing*	*Relocation*	*Well-being*

Previous studies were cited that assessed before and after effects
of relocation upon the social and psychological well-being of the elderly. These studies found an association between levels of residents'
well-being and relocation to a new housing environment, although comparisons with non-moving groups were cross-sectional rather than longitudinal. The present study investigated the impact of rehousing and the
well-being of groups of elderly in a longitudinal comparison-group
design. This design enabled subjects to be investigated regarding
levels of change that were experienced following a move.

Applicants for senior housing (N = 574) and elderly community residents (N = 324) were interviewed at one point in time; 591 were interviewed one year later. Five new housing sites were assessed in terms of
change experienced by residents during the first year of occupancy as
compared to change in groups of community residents who did not relocate. Housing sites consisted of two low-rent public housing sites and
three federally assisted projects (all highrise buildings) for persons
aged 62 or older. Change over time among the rehoused was compared to
that among the comparison subjects by means of multiple regression analysis which controlled for original state of well-being, demographic
characteristics, and the initial health state of subjects.

The rehoused were significantly better off than the community residents on five factor derived indices, poorer in functional health, and
not different according to three other indices. The size of the favorable effect was small, but found to be relatively stable and was not
emphasized as possessing great magnitude by the authors (due to restrictions cited in the sampling techniques), who stated that even the best
housing environment cannot be expected to totally solve immediate or far
reaching problems of the tenants.

Some implications of the findings were: 1) rehousing the elderly
does have definite advantages as ascertained by tenants in that positive
effects were evident when studying change over a period of one year; 2)
self-perceived change for the better was the most pronounced indicator
of a positive housing effect; 3) rehoused persons were more involved in
social activities outside the housing site per se than were nonrehoused
subjects; and 4) a substantial decline in functional health status of
rehoused subjects was identified. This finding contrasted with the general improvement in well-being found in other measures. Several possible explanations for the less healthy status of the relocated persons
were postulated. The authors concluded that, with exception of the
health variable, the findings were very consistent with Carp's (1966)
findings at Victoria Plaza. Although the physical health of relocated
residents in the current study diminished over a year's time, certain
indicators of well-being showed a positive change during this same
period.

"Physical Structure and the Behavior of Senile Patients Following Ward Remodeling," *International Journal of Aging and Human Development*, Vol. 1, No. 3, July, 1970, pp. 231-239.

KEY WORDS:

Man	Setting	Relational Concepts	Outcomes
Senile elderly	Geriatric center	Remodeling/ behavior	Behavioral changes

   This study examined the behavioral changes among senile elderly patients of a geriatric center before and after ward remodeling in which they were staying. The rooms which were altered were located in a 48-bed wing of a multiuse geriatric center. The wing's patients were primarily diagnosed as chronic brain syndrome (senile brain disease or arterial arteriosclerosis), moderate to severe. Most were ambulatory, but some required a wheel chair. The original structure's decor was institutional, and there was little to break the visual monotony, as well as the behavioral monotony. The appearance was essentially drab with bedspreads, window shades, and walls in grayed color, all lacking pattern and design. The superficial look of this wing was distinctly that of a deprived environment.

   The area selected for alteration consisted of two large rooms, one occupied by four and the other by five female patients. The consulting architects (Bertrom Berenson and Tidewater Design Associates), in collaboration with the institution's staff, redesigned the area so as to achieve the following goals: 1) provide more private space for each patient; 2) provide space for small social group interaction; 3) improve open communication between private, small social and larger social (i.e., the corridor) space; 4) enhance personal identity through provision of mirror, dresser, and shelf space for personal belongings; and 5) enrich the environment using bright and contrasting drapes, wall and floor coloring, textures, lighting, and furnishings such as a planter, pictures, and a bird cage.

   In relocating the patients, two of the nine patients died, and a third was hospitalized before remodeling was begun. It was observed that relocation was disturbing to the patient's routine. In addition to the emotional disturbance attendant to the move, three of the six relocated patients died within six months following their move. An important question which must be considered by this study is whether this temporary displacement may have accelerated the deaths.

   A direct observational method was used by which to document change. A total of nine patients were observed in the old setting and six in the new setting. As mentioned, all patients were diagnosed as suffering from organic brain syndrome. Their ages ranged from 72 to 86, and their period of institutionalization from one to eight years.

   The behavioral observations appeared to indicate a decrease in personal interactions, a possible decrease in self-maintaining behavior, and an increase in active interest which is unrelated to the change in physical structure. Private space was provided and utilized by the patients to varying degrees. Two patients were never in their own rooms, while another was almost always there.

Even though a space was built for interaction, the interaction
decreased. The reason given for this was that the patient was no longer
close to at least 3 or 4 other people 100 percent of their day. One
person had in effect decreased the probability of interacting by choos-
ing to be alone a portion of the day, a choice not as easily achieved in
the original setting.

Interaction patterns with staff showed that service contacts did
not change. The number of social interactions associated with the same
staff members decreased. Following remodeling a fleeting glance from
the main corridor was sufficient to see if everything was alright, as
opposed to the original setting where staff had to enter directly into
the rooms. Expansion of range and the amount of ambulation was the most
clearly achieved goal of the architecture. Giving a choice and opening
up barriers greatly increased the amount of movement and exposed the
patient to a greater variety of auditory, visual, kinesthetic, and
social stimuli.

The fourth and fifth goals, enhancement of personal identity, and
enrichment were approached only peripherally. As the authors noted, a
more focused experimental approach is required to determine to what
extent these two goals may be achieved. Nevertheless, the decrease
noted in incidence of self-maintaining behavior suggested that the provision of more personalized space and objects does not lead automatically to any increase in general self-care activity.

LAWTON, M.P., M.H. KLEBAN, AND M. SINGER   145

"The Aged Jewish Person and the Slum Environment," _Journal of Gerontology,_ Vol. 26, No. 2, April, 1971, pp. 231-239.

KEY WORDS:

Man	Setting	Relational Concepts	Outcomes
Elderly	_Inner city_	_Housing, social,_	_Deprivations_
Jewish	_neighborhoods_	_economic conditions_	

Elderly Jewish residents of an older urban area were interviewed to
assess their present needs, circumstances, and how they compared with
other elderly persons. Previous studies were cited which sought to provide better understanding concerning the plight of the older person in
the city with regard to interactions between environment (housing and
neighorhood) and elderly persons' economic, health, and social
problems.

The site of the study was an inner city neighborhood in Philadel-
phia which had undergone a virtual transformation within recent decades.
The non-black percentage of residents dropped from 84% (1940) to less
than 1% (1970). A detailed neighborhood profile was presented. The
interview sample consisted of 115 persons (71 females and 44 males).
Questions focused on a wide array of content areas including housing
needs, types of desired services, environmental surroundings, and a set
of detailed background questions. The use of content area questions
similar to those employed in a previous study by one of the authors

enabled a comparative look at this elderly group in relation to other
groups of aged persons in other communities residing in specialized
housing for the elderly. Subject groups from this previous study were
chosen for the comparative analysis. These groups (SM, JC, AL and CL)
were comprised of persons who lived in other urban neighborhoods that
were of different demographic mixture.

The data were categorized and presented in eight distinct areas
that were referred to as indicators of well-being. The four groups of
elderly persons were compared according to these major variables: 1)
health, 2) mobility, 3) functional independence, 4) family relationships, 5) peer interaction, 6) leisure-time activity, 7) morale, and 8)
housing satisfaction.

The results of analysis for the first three indicators of well-being depicted a principal subject group member (SM) as being in poorer
health than peers in other settings and especially restricted in physical mobility. However, this group performed most activities necessary
for everyday living without measurable impairment. For indicator number
four, the level of family interaction was about equal for the four
groups. For the very few SM members with even one friend, indicators of
peer interaction showed that 23% of the SM residents visited these
neighbors every day, compared to only 8% in group JC. It appeared that
SM residents were socially dissatisfied and deprived, in that their
level of interaction with friend and neighbors was lower than for two
other comparison groups. The SM group partook in significantly fewer
leisure-time activities than other group members. A uniform test (PGC
Morale Scale) showed the SM group as having the lowest morale of all
groups. Also SM residents were extremely dissatisfied with physical
housing conditions.

The authors summarized that the SM group was: 1) deprived in economic status, 2) in poorer health, 3) was less mobile and interactive
with peers (and possibly with family, and 4) was less satisfied with
housing than were several other groups of elderly people. They performed
self-maintaining activities as frequently as did the other groups. The
strains to perform daily activities in spite of several types of deprivation were reflected in SM members' lower morale. Causes of their
deprived status were discussed in terms of direct environmental influences, selected migratory patterns, and unfortunate circumstances
intrinsic to the aging process.

LAWTON, M.P.   146

"The Relative Impact of Congregate and Traditional Housing on Elderly
Tenants," *The Gerontologist,* Vol. 16, No. 3, June, 1976, pp. 237-242.

KEY WORDS:

Man	Setting	Relational Concepts	Outcomes
Elderly	Housing	Housing arrange-ments	Satisfaction

A great deal of emphasis in public supported housing for the elderly has been placed upon the provision of basic services such as eating
and socializing within the housing complex. Some social scientists have
argued that this reliance on "congregate" types of housing discourages
the normal social patterns found in traditional, independent living
arrangements. This article attempted to measure the changes in well-being that might be ascribed to each of these two environments.

Interviews were conducted with 122 residents of a housing development that provided internal food service and organized activities and 78
occupants of self-contained apartments before occupying and one year
following occupancy of these types of housing. Both populations were of
similar ethnic background, although the traditional housing residents
tended to be slightly younger with higher monthly incomes. Nine categories of well-being were defined in terms of leisure patterns, morale,
sociability, child orientation, housing satisfaction, change, external
involvement in society, activities, and satisfaction with the status
quo. Hierarchical multiple regression analyses were used to determine
the effect of living in traditional versus congregate housing.

The author concluded that the congregate form of housing had the
disadvantage of discouraging external contact. Although the congregate
housing residents showed improved ratings in morale, housing satisfaction, and closeness of internal social networks, the provision of a man-
datory food service program and organized social events tended to diminish the independence and overall mobility of the resident compared to
one living in the more traditional form of housing.

LEADERER, B.P., R.T. ZAGRANISKI, AND J.A.J. STOLVIJK   147

"Estimates of Health Benefits Due to Reductions in Ambient NO_2 Levels,"
Environmental Management, Vol. 1, No. 1, 1976, pp. 31-37.

KEY WORDS:

Man	Setting	Relational Concepts	Outcomes
Human	United States	Nitrogen dioxide pollution	Health

The 1970 Air Quality Act has established that the nitrogen dioxide
(NO_2) concentration in the ambient air shall not exceed 0.05 parts per
million (ppm). Based upon measurements of the effects of NO_2 on a population, the authors presented a model that predicted health benefits on
the population of the United States by a reduction of the NO_2 level
below the 0.05 ppm standard in all areas that presently are above.

Using the 1969 Chattanooga Study which measured the respiratory
effects of NO_2 on the general population over a 24 week period, the
researchers calibrated a mathematical model which predicted the effects
of exposure at levels varying betweeen 0.04 and 0.11 ppm. They then
took measured levels of NO_2 from cities across the country from 1960 and
1970 census information and projected an expected level for the year
1973. A simulation was then performed in which actual pollution levels
were injected into the model and graphs of NO_2 levels versus health consequences were plotted.

The study predicted that there were 43.9 million respiratory illnesses due to NO_2 exposure during 1973, of which 3.7 million occurred in
Los Angeles alone. Illnesses due to exposure at the 0.05 ppm level or
above accounted for 33.2 million of the total figure with approximately
26.5 million occurring in the eastern United States. "It revealed that
a 50 to 60 percent reduction in NO_2 concentrations would essentially
eliminate excess illnesses of acute respiratory disease, or achieve a
100 percent benefit." Correlations were also postulated concerning the
health benefits in relation to proposed automobile emission standards,
and it was found that "an emission standard of 2.0 grams of NO_2 per mile
or less would be necessary to prevent any excess respiratory illnesses
due to ambient NO_2 concentrations (that stationary sources are controlled to the same extent)."

LECUYER, R.   148

"Social Organization and Spatial Organization," _Human Relations,_ Vol.
29, No. 11, November, 1976, pp. 1045-1060.

KEY WORDS:

| | | Relational | |
Man	Setting	Concepts	Outcomes
Human	Tables	Spatial organi-	Social interaction
		zation	

This study attempted to answer the question of what are the relationships between spatial arrangement and social interaction. The article investigated the hypothesis that spatial configuration is a primary
factor in resulting human behavior and the reverse thesis that social
interaction is a determinant of the ways people arrange themselves in a
physical environment.

The first experiment divided two groups of 80 French psychology
students into subgroups of five each to perform a task at either a rectangular or circular table. The hypothesis of this experiment was that
by observing the interpersonal behavior of these groups, the experimenter could predict that the subjects at the rectangular table would
rely on a leader-type method of problem solution while those at the circular table would be much more free to communicate on a common level.
This would confirm the notion that physical setting determines a group's
behavior. In the second experiment, 80 male French factory foremen were
divided into groups of fours and seated at a circular table with the
experimenter occupying a random seat. After a task had been defined by
the experimenter, an unexpected change was made to a room with a rectangular table with a head chair. By comparing the seating arrangements
before and after the change, the author hoped to prove the hypothesis
that social organization is a determinant in the arrangement of seating
patterns.

The results of the two experiments confirmed the two reciprocal
hypotheses of the author but also produced some surprising results. In
the first experiment, although the expected leadership pattern emerged
from the rectangular arrangement, there were no significant increases in
communication patterns at the circular table. It appeared, therefore,

at the small group size the physical arrangement was not a significant factor. In the second experiment, even though the experimenter was the obvious leader of the task problem, the occupany of the head chair after the move to the rectangular table was not usually reserved for him as was expected by the author. Instead, the leadership patterns that emerged from the four foremen were reflected in the new seating arrangement, with the experimenter being relegated a minor seating position.

The author concluded with two generalizations. First, people do not choose their seating arrangements at a table at random, but rely on a specific social organization. Second, whether the spatial configuration is chosen by authority or by the collective wisdom of the group, an appropriate relationship between space and social relations is maintained.

LENNQUIST, S., P.O. GRANBERG, AND B. WEDIN   149

"Fluid Balance and Physical Work Capacity in Humans Exposed to Cold," _Archives of Environmental Health,_ Vol. 29, No. 5, November, 1974, pp. 241-249.

KEY WORDS:

Man	Setting	Relational Concepts	Outcomes
College students	_Thermal environment_	_Thermal exposure_	_Physiological changes_

Earlier reports have indicated that exposure to cold air results in a greater urine flow than would otherwise occur. The build-up of a strong negative water balance could have implications regarding the capacity of the individual to do physical work. In order to study this possible relationship, healthy persons were studied in climate-chamber experiments under carefully controlled conditions.

Twenty-six healthy students each spent 24 hours immediately preceding the testing in a specially constructed climate chamber while it was set at 28°C. The subjects were briefly moved to a neighboring room also set at 28°C until the original climate chamber had its temperature stabilized at 15°C. Dressed in bathing trunks, socks, and shoes the subjects were returned to the chamber for periods varying from one to five days. Care was taken to maintain the normal 24 hour rhythm of the individuals. Activities were standardized and consisted of exercises such as ergometer bicycling and walking up and down a staircase in the chamber. Rest periods, meal times, and measurements were also taken at prescribed times.

Several physiological changes were noted during the course of the study: 1) rectal temperatures fell a mean value of 0.5°C; 2) there was a significant fall in the pulse rate; 3) body weight decreased on the average 1.6%; 4) blood volume fell approximately 1 liter; 5) urine flow increased significantly, and 6) excretion of sodium, chloride, and calcium also increased significantly. In sum, "even under moderate cold there occurred an increase in urine flow to a greater extent than could be explained by reduced insensible water loss, leading to a negative

water balance, orthostatism, and reduced physical work capacity." The
authors hypothezied that calcium might influence renal responsiveness
and be the trigger for cold-induced diuresis.

LEFF, H.L., L.R. GORDON, AND J.G. FERGUSON   150

"Cognitive Set and Environmental Awareness," *Environment and Behavior,*
Vol. 6, No. 4, December, 1974, pp. 395-447.

KEY WORDS:

Man	Setting	Relational Concepts	Outcomes
Students	*Street views*	*Cognitive sets*	*Perception*

Individuals were conceptualized as being able to exert considerable
control over their perceptual experience and environmental awareness
through the deliberate variation of their cognitive sets. Cognitive
sets were defined as mental plans for selecting and processing information. This topic was explored in light of several lines of research and
theory. This paper focused on a project concerning the effects of
selected cognitive sets. Eleven experiments were conducted to examine
such sets as concentration on seeing the environment as a collection of
abstract forms and concentration on imagining how man might make the
environment more pleasant.

The following five psychological considerations were viewed as critical in the design or evaluation of a cognitive set with respect to
sets' potential for influencing perception and awareness of environmental issues: 1) competence; 2) comprehension; 3) complexity; 4) composition; and 5) adaptation (comparison) level. Two major types of experimental design were employed. In one of the experiments seeking between-
subjects variation (Experiment 3), sixty subjects participated, where 20
individuals were randomly assigned to one of three set conditions. A
slide of a street scene was presented and subject groups one, two and
three each received different sets of instructions for responding to the
stimulus. A series of 23 semantic differential scales followed. Contents of the bi-polar scales were provided at the outset. Finally,
dependent variables consisting of a set of background information questions and tasks were obtained from individual subjects.

The results indicated that: 1) the set to view a scene as an
abstract collection of shapes, lines, textures, and colors increases the
judged complexity of the scene and was rated as more interesting, yet
more difficult than "normal" viewing; 2) the cognitive sets to evaluate
or to figure out the values represented by the human influences in a
scene tended to increase awareness of environmental problems but were
relatively unenjoyable and difficult, although these aspects were not
regarded as uninteresting, and 3) the cognitive set to imagine how man
might increase the pleasantness of a given scene decreases the judged
pleasantness of the actual sense of perceptual control. This was found
to be relatively easy, enjoyable, and interesting when both viewed abso-
lutely and when compared to normal viewing. This set was also more
effectively activated when occurring later within a series of different
cognitive sets than when occurring earlier in the sets. The results

were discussed in terms of theoretical concepts outlined in the introduction and with brief reference to specific educational implications.

LEVINE, R.   151

"Occupational Lead Posioning, Animal Deaths, and Environmental Contamination at a Scrap Smelter," *American Journal of Public Health*, Vol. 66,
No. 6, June, 1976, pp. 548-552.

KEY WORDS:

Man	Setting	Relational Concepts	Outcomes
Employees neighboring residents	Scrap smelter	Occupational lead poisoning/envi-ronmental contami-nation	Health hazard

This study described a severe episode of occupational poisoning and
environmental contamination at a lead recovery plant in Alabama. Thirty
of the thirty-seven employees had blood levels indicating unacceptable
lead absorption. Eight current and previous employees had been hospitalized with lead colic, and another with encephalopathy. Levels of lead
in surface soil and vegetation at the smelter were high and decreased
with distance. Animals on nearby pastures had died and lead levels in
the blood, milk, and hair of large and small animals in the area were
elevated. Adults living within 100 meters of the smelter had higher
blood and hair levels than controls who lived at a greater distance, but
there was no evidence of lead toxicity.

The human, environmental, and animal data collected in this investigation implicated the lead scrap smelter in Troy, Alabama, as a major
local source of lead contamination. The temporal distribution of the
human cases and of the animal deaths suggests that the severity of the
contamination must have increased sharply after 1970, coincidental with
increases in smelter production.

Many of the features of this episode have been recorded previously
in the areas surrounding other primary and secondary lead smelters.
Occupational poisoning, contamination of air and soil, animal deaths,
and absorption of lead by members of nearby communities, particularly
children, appear to be rather common occurences in such locales.

LINN, M.W., EM. CAFFEY JR., C.J. KLETT, AND G. HOGARTY   152

"Hospital vs. Community (Foster) Care for Psychiatric Patients,"
Archives of General Psychiatry, Vol. 34, No. 1, January, 1977, pp.
78-83.

KEY WORDS:

Man	Setting	Relational Concepts	Outcomes
Psychiatric patients	Hospital, foster care homes	Patient self-management	Adjustment

Foster care homes have been used to provide an opportunity for those patients with various psychiatric disorders to function adequately in the community, but who would not best be served by placement back with their own family. The value of such a treatment procedure has been the object of a great deal of uncertainty. Foster homes, for instance, are often criticized for only providing meager custodial services which in many cases tend to be worse than the continued hospitalization environment they are trying to replace. The aim of this study was thus to compare the effectiveness of a hospital treatment setting to the results of foster care placement.

Five hundred and seventy-two Veterans Hospital patients were randomly selected from five hospitals to be assigned to foster care homes or continued hospitalization. All were male, had a psychiatric disorder, and had no previous experience with foster care. In terms of diagnoses, "71% were diagnosed as having schizophrenia, 17% had chronic brain syndrome, and 33% had a diagnosis of alcoholism, either currently or in the past." Patients were admitted according to a staggered scheduling procedure which allowed the hospitals to prepare the patients according to their usual time frame. They were examined before assignment and monitored while at the placement setting. Evaluation of the effectiveness of the programs was done four months later on the basis of functioning, mood, activity, and overall adjustment. Four measures were used, and assessments made by people who knew the patient, but who had no particular investment in the success of foster care as a hospital program.

The results revealed that although little change was found in the social functioning of the patient as a result of hospital preparation for foster care, within four months of being placed in a foster care setting, subjects showed significant improvements in terms of overall adjustment. These findings were consistent across all of the hospitals studied, leading the authors to conclude that, "findings suggest that attention should be given to selection criteria, that lengthy preparation may be unnecessary, and that foster care is superior to hospitalization for patients who cannot return to their homes."

LIPMAN, A. AND R. SLATER   153

"Status and Spatial Appropriation in Eight Homes for Old People," *The Gerontologist,* Vol. 17, No. 3, June, 1977, pp. 250-255.

KEY WORDS:

Man	Setting	Relational Concepts	Outcomes
Elderly	*Homes for the elderly*	*Seating arrangements*	*Spatial & social organization*

The hypothesis presented within this article was that communal seating arrangements in homes for the elderly are determined and reinforced by the social status of a person (given his or her rationality and sex). It is postulated that physical arrangements of space can be categorized as being "good" or "bad" and that rational, male occupants of homes will predominate in the selection of the better seating arrangements.

Eight homes for the elderly in England and Wales were used as sample environments in this study, each chosen on a number of physical criteria related to the overall research design. Community seating arrangements were rated according to their proximity to the television, relation to exterior views, and closeness to dining facilities. The seating arrangements tended to be static and maintained in their original configurations by the staffs. The occupants were categorized into various stages of rationality using standard tests as well as the subjective opinions of the staff members. Thirty-four communal sitting spaces were studied using field logs and structured observation techques.

The study showed that the more rational occupants of the homes dominated the better sitting areas and seating arrangements. This conclusion was supported by the fact that more rational occupants tended to view the seats as their territory, as well as the fact that staff members placed irrational occupants in seating not taken or claimed by more lucid residents. The hypothesis that male occupants were more likely to have access to better seating arrangements was not supported by the study. The authors concluded that 1) the methods of scoring the seating patterns were insufficiently discriminating, 2) spatial appropriation is marginal in homes for the elderly, and 3) spatial appropriation is determined primarily by staff members, with little individual choice being given to the occupants themselves.

LUDWIG, A.M.   154

"Self-Regulation of the Sensory Environment," *Archives of General Psychiatry,* Vol. 25, No. 5, November, 1971, pp. 413-418.

KEY WORDS:

Man	Setting	Relational Concepts	Outcomes
Human	*Laboratory*	*Sensory deprivation, sensory overload*	*Environmental preferences*

A review of the literature concerning the manner with which man reacts to his sensory environment revealed that virtually all of the existing research has concerned itself with sensory deprivation. The common methodology utilized has tended to revolve around the reduction of sensory input or the exposure of subjects to repetitive, monotonous stimuli. Citing the need for studies which expose people to "sensory overload," the author undertook an experiment which compared sensory deprivation and sensory overload conditions to "normal" sensory input. Since the research was still in progress at the time the article was written, it was the author's intention to "discuss some of the kinds of data this paradigm can yield as well as some preliminary findings," rather than representing any definitive conclusions.

Subjects were exposed to three $2\frac{1}{2}$ hour test sessions: 1) sensory deprivation (SP), 2) sensory overload (SO), and 3) normal sensory input (NS). Lighting was randomly altered by means of a light cycling programmer which automatically changed the intensity and color of the

lights in a test chamber. A Moog Synthesizer created a wide array of
unusual and unpatterned sound which was ultimately recorded in order to
provide each subject with the same set of circumstances. The only difference between the NS and SO conditions was the intensity of the sound.
The master control panel allowed the conditions to remain constant
across test sessions.

Upon entering the test chamber, each subject was seated in a comfortable recliner chair in front of a panel containing four buttons.
Two of the buttons enabled the subject to control the sensory input,
while the other two were nonoperative "dummy" buttons. The purpose of
the nonoperative buttons was to measure general button-pressing activity
rather than goal-directed activity. Included in the panel was a "panic"
button which would immediately bring the room back to normal lighting
and sound. In the event that the panic button was pushed, an experimenter would immediately speak with the subject regarding the cause of
the distress. If the subject wished to continue, a reset button was
pushed and the rest of the conditions were presented. If the subject
did not wish to continue the experiment was terminated at that point.
The behavior of the subjects was monitored from the control room by
means of a closed circuit television camera.

Results indicated that subjects worked much harder to avoid sensory
overload than they did to alter sensory deprivation. Under sensory
deprivation conditions there was no marked preference for either light
or sound. In sensory overload situations, however, subject responses
indicated that they were far more affected by sound than by lighting.
Emphasizing that the results are preliminary, Ludwig believed that "further study in this area will yield important clinical insights about the
behavior of man and suggest innovative ways for modifying that behavior
should the need arise."

LUNDEN, G.   155

"Environment Problems of Office Workers," *Build International,* Vol. 5,
No. 2, March/April, 1972, pp. 90-93.

KEY WORDS:

Man	Setting	Relational Concepts	Outcomes
Office workers	*Office buildings, Sweden*	*Work environments*	*Job satisfaction*

Nine office buildings in Sweden were selected to subjectively study
how office workers feel about their surroundings. The buildings were
selected as being representative of a wide range of climates, as well as
possessing diverse design techniques for dealing with those climates.
The main focus of the investigation was to rank a number of factors concerning contentment in an office setting and relate them to the actual
physical surroundings.

In each building between 20 and 40 rooms were chosen at random,
with a total of 450 people working in these rooms being interviewed.

Each person was given nine cards, each illustrating a contentment factor. The task was to rank each of the cards in order of importance for
job satisfaction. The subjects were asked to make their choices based
on offices in general, rather than limiting their decisions to their own
jobs. The results of the ranking was as follows: 1) type of work, 2)
the chief executives and the salary, these being rated equally, 4) col-
leagues, 5) chances of promotion, 6) security of employment, 7) the
office environment, 8) working hours, and 9) location of the place of
work. Hence the actual office environment did not reveal itself to be
of major importance.

As a further investigation of the physical setting, the workers
were asked to judge ten factors connected with the daily environment and
working conditions in the offices. One major finding was the strong
desire to open windows due to the heat. In the two buildings in which
it was not possible to open the windows, 90% of the respondents wished
that they could be opened. In fact, the wish to have the windows open
was so strong that bad air from the outside was not seen to be a
deterrant.

Other preferences centered around individuals being able to have
their own room, with the opportunity to shut the door so as not to be
disturbed. The present heights of the rooms were generally satisfactory, but if the widths were between 2 and 4 meters many respondents
thought them to be too narrow. No person expressed this opinion for
rooms with 4-6 m. widths. Although the environmental conditions did not
rank as high as some other factors in terms of job satisfaction, certain
design features were deemed to be relatively important among the workers
surveyed.

MACDONALD, W.A. AND E.R. HOFFMAN   156

"The Recognition of Road Pavement Messages," _Journal of Applied Psychology,_ Vol. 57, No. 3, June, 1973, pp. 314-319.

KEY WORDS:

Man	Setting	Relational Concepts	Outcomes
Students	_Actual & simu-lated roadways_	_Threshold recognition_	_Roadway lettering effectiveness formulas_

The relationship between recognition threshold and degree of elon-
elongation of letters used in road pavement messages was investigated.
The central hypothesis is that the degree of letter elongation should be
kept to a minimum because it is economically unnecessary to expend ener-
gies painting extremely large road pavement messages and because there
is evidence that from a safety standpoint, a sizable number of commonly
used paints have lower coefficients of friction than an unpainted road
surface - with a subsequent rise in the risk of skidding. The research
approach was to establish the form of the relationship between recognition threshold and vertical visual angle subtended for letters of varying elongation.

The relationship between the fixed value of visual angle and the point where recognition takes place was studied in addition to several complicating factors, whose effects would be to modify the operation of the law of the visual angle so that the advantage of greatly elongated letters would decrease further. Experiments were conducted in the field and in a laboratory setting. Twelve undergraduate students comprised the subject group. Laboratory performance was recorded in terms of threshold angle of tilt, while filed performance was recorded in terms of threshold distance. Both measures were converted to threshold vertical visual angle subtended by the letter.

It was found that in both situations the normally proportioned letters were recognized at smaller visual angles than the more elongated letters, and that increases in letter elongation did not produce increases in recognition distance directly proportional to the increases in the vertical visual angle subtended. Mathematical models based on the relationship between perceived and real distance largely describe the observed effect, and a formula is given by which traffic engineers can calculate the necessary degree of letter elongation for a desired recognition distance on a roadway surface.

MAITREYA, V.K.   157

"Daytime Artificial Lighting in Office Buildings in India," *Building and Environment,* Vol. 12, No. 3, 1977, pp. 159-263.

KEY WORDS:

Man	Setting	Relational Concepts	Outcomes
Male	*Buildings, India*	*Gloom, lighting preference, performance*	*Artificial lighting standards*

The rising costs of land in India have resulted in buildings being built with low ceilings and a high density population. Under those design constraints it is not easy to provide the recommended lighting levels by natural means alone. In addition, the artificial lighting used during the daylight hours often does not satisfy the minimum requirements, which may affect work performance. This study was designed to examine the standards necessary for office tasks.

A field study was undertaken to determine the amount of artificial light which was in actual use during the daytime. Measurements of various buildings were taken during the months of June and November under "stable sky conditions." The variants studied were room sizes; window size, location and orientation; louvres; and artificial light sources.

Analysis of the data revealed that the amount of light required for satisfactory performance of an office task varies between 100-200 lx. In a tropical country such as India, thermal comfort limits the size of the window apertures. It was found that in many cases the part of the room away from the windows presented a gloomy environment.

Based on this knowledge a subjective laboratory experiment was conducted in order to suggest artificial lighting standards. The window
apertures of a controlled room could be altered from 2.0 to 13.0% of the
floor area. Seven male subjects between 25-35 years of age were asked
to make subjective judgments of the lighting while performing a reading
task. The task consisted of reading a paper containing a number of
spelling mistakes, so that changes in performance could be compared with
the amount of illumination present.

Results revealed an overall improvement in the time taken for the
test and the number of mistakes committed with an increase in lighting
levels. It was found that, "an average value of 150 lx task illuminance
with modelling vector between 1.5 and 2.5 represents an acceptable lumi-
nous environment. The task lighting of such an interior is sufficient
and the directional qualities give a pleasant environment indoors."

MARGULIS, S.T.   158

"Conceptions of Privacy: Current Status and Next Steps," *Journal of
Social Issues,* Vol. 33, No. 3, 1977, pp. 5-20.

KEY WORDS:

Man	Setting	Relational Concepts	Outcomes
Diverse user groups	*Recent studies*	*Privacy*	*Summary of current privacy concepts*

Margulis identifies three stages in the development of a theory of
privacy. These include: 1) studies, observations and cases that demonstrate the importance and viability of a behavior concept of privacy; 2)
systematic explorations of a conceptual definition of privacy; and 3)
systematic exploration of the whys and hows of privacy-behavioral
analysis.

Privacy has had a variety of meanings. Common meanings of privacy
include separation from others through control over information, space
or access, including simply being or working alone. Empirical meanings
of privacy include patterns of involvement and interaction and the role
of environmental mechanisms for controlling interaction. Legal meanings
of privacy include personal control over personal information including
environmental intrusions such as noise.

Privacy is defined as "a whole or in part the control of transactions between person(s) and other(s), the ultimate aim of which is to
enhance autonomy and/or to minimize vulnerability."

Altman's ecological, or social systems analysis is examined as a
movement toward the third stage. This analysis of privacy emphasizes:
1) multiple levels of behavior, 2) behavior operating as a coherent
system, 3) a dynamic view of behavior, 4) a dialectic view of behavior,
and 5) a view of the environment as a determinant of and as an extension
of behavior.

The strengths of this analysis lie in the comprehensive view taken and in its socioenvironmental orientation. Problems arise from the use of concepts such as "balance," control and self. The transactional components of privacy equated with social interaction can potentally create too wide a view of privacy, which may lead to ambiguity.

Theorists do not agree on whether privacy is a behavior, attitude, process, goal, or phenomenal state. Altman's selective control over access to the self is contrasted to Kelvin's perceived negation of the power of others. At the same time Pastalan's concept of privacy as a fundamental form of human territoriality is contrasted with Altman's human territoriality as a mechanism for obtaining desired levels of privacy. These theories demonstrate the ties between environment, architecture and the behavior analysis of privacy.

MARKSON, E. AND J.H. CUMMING   159

"A Strategy of Necessary Mass Transfer and Its Impact on Patient Mortality," _Journal of Gerontology,_ Vol. 29, No. 3, May, 1974, pp. 315-321.

KEY WORDS:

Man	Setting	Relational Concepts	Outcomes
Elderly	_Hospitals_	_Involuntary relocation_	_Mortality_

This study focused on the impact of mass tranfer upon the mortality rates of 2,174 patients whose relocation was a portion of a plan that dealt with significant budget reductions mandated for the New York State Department of Mental Hygiene in April, 1971. These psychiatric patients were transferred to different hospitals and the mortality rates were monitored during an 11 month period after the relocation.

A detailed description of the department plan to deal with budget reduction was included. As a result of the implementation of the plan to cope with reduced funding, three types of institutions were developed: 1) those who lost patients by transfer, had staff dismissed, and ended with a better patient-staff ratio; 2) those who received patients by transfer with no additional staff (and had a worse patient-staff ratio); and 3) those who neither received nor lost patients (with a constant patient-staff ratio). The majority of the relocated patients observed in the study possessed chronic mental disorders.

Less than 10% of the elderly transfer patients died during the 11 month period following relocation. The mean age of the geriatirc patient group (N = 494) transferred was 74.97. Ages ranged from 65 to 95. Elderly transfer patients were healthier than all chronic geriatric patients, including both physically sick and well elderly.

Results suggested being life-threatening to the chronic or elderly patient, no evidence was found here which supported the claim that relocation had a deleterious effect on relatively physically able patients. The change in environmental setting, from a less active to a more active

environment, did not have a noticeable effect on the proportion of relocated patients who died. There was no overall relationship between
cause of death and date of death for the 36 males and 50 females who
died during the 11 month period.

The author concluded that there was no evidence that relocation of
this group of mostly schizophrenic patients was stressful to the point
of affecting the mortality rate. Despite substantial budget cutbacks by
the state governing agency which resulted in the phasing out of some
wards and overcrowding in others, the chances for survival for most of
the patients in this study did not decrease following involuntary relocation.

MARKUS, E., M. BLENKNER, M. BLOOM, AND T. DOWNS   160

"Some Factors and Their Association with Post-Relocation Mortality Among
Institutionalized Aged Persons," *Journal of Gerontology,* Vol. 27, No. 3,
July, 1972, pp. 376-382.

KEY WORDS:

Man	Setting	Relational Concepts	Outcomes
Two elderly groups	*Institution*	*Relocation, phy- sical status, mortality, coping, mental status*	*Undifferentiated mortality rates*

Prior studies have suggested that the involuntary relocation of the
elderly has either directly or indirectly been linked to increased mortality and morbidity for some groups and decreased mortality for others.
Most of these studies have focused on age and sex, which the present
authors regard as insufficient if not supplemented by other factors. In
this study, relocation was conceptualized as consisting of two separate
stages: 1) deprivation of familiar cues and environmental supports; and
2) coping with new sets of stimuli in an unfamiliar environment. It was
hypothesized that the field approach and persons' mental status were two
factors that were directly involved in studying differential effects of
relocation.

The entire populations of two homes for the aged (105 males and 268
females) were tested for perceptual fixed-dependence, mental status, and
physical status. Subjects were then relocated from old downtown buildings to new suburban facilities. Pre- and post-relocation interviews
were conducted with each subject. Data on mortality were gathered for
nine months following relocation. The three major variables were opera-
tionalized via a series of standardized questionnaires. Within the
third major variable (physical status), seven indices were employed: 1)
an index of the degree of care assigned to a particular resident; 2)
physician ratings on the degree of organic brain damage; 3) staff
assessments of mortality; 4) a physical functioning questionnaire; 5) an
index that ascertained the degree of independence-dependence of the
residents in routine tasks; 6) an index of morbidity that monitored
residents' days ill in bed; and 7) a self-reported scale that monitored
persons degree of ambulation. Mean ages were 81.2 and 79.7 years for

males, 79.7 and 80.7 years for females in subject groups one (JOHA) and
two (HMA), respectively. Both groups were highly comparable along a
series of demographic characteristics.

The association between the tested factors and post-relocation mortality observed in the first home was not reproduceable in the second.
Combinations of factors were studied through discrimination analysis and
the most accurate and illuminating predictive model dealt with sex factors. Of the major factors used in the analysis, very little discrimination was detected between survivors and nonsurvivors. These four factors were age, MSQ (Mental Status Questionnaire), ward (care index), and
CEFT (Childrens Embedded Figures Test). The results indicated that the
imposition of a new and unfamiliar environment was confounded with low
mental status. At JOHA no residents who scored 4 or more on the CEFT
died, while at HMA, five residents who scored above 4 did expire.

The authors attempted to explain the failure to reproduce the JOHA
findings via the possibility that other, more "elusive factors" such as
"readiness for change," the "will to live," and "tolerance for stress"
might need to be studied. The question was posed "Is there some threshold beyond which change becomes more stressful?" They concluded that the
task of analyzing the relocation of institutionalized elderly persons
and subsequent mortality rates is an open-ended problem that requires
additional study.

MARSHALL, N.J.   161

"Privacy and Environment," *Human Ecology,* Vol. 1, No. 2, September,
1972, pp. 93-110.

KEY WORDS:

| | | Relational | |
Man	Setting	Concepts	Outcomes
Students,	*Suburbs*	*Environmental*	*Perceived*
parents		*surroundings*	*privacy*

The relationship between environmental surroundings and a person's
perceived degree of privacy is the basis of this study. The author suggests that by obtaining orientations about privacy from a sample population she can then predict subsequent attitudes that this population
possess concerning their past and present physical environments.

One hundred and forty-nine college students and a sample of 101 of
their parents were used as the experimental population for this study.
Socioeconomic indices were established and a wide range of social levels
(from professional to unskilled labor) were found to exist within this
sample. The population was administered a questionnaire to determine
specific attitudes about ways of controlling the degree of privacy in
terms of such devices as visual and auditory barriers in the home, density of home and neighborhood, and frequency of contact with neighbors
and strangers. A second questionnaire was used to determine past environmental experiences and present housing configurations of the sample
in order to correlate attitudes with specific physical surroundings.

Several specific correlations between privacy and environment emerged from the study. Subjects judging their present home as crowded were found to have more persons per room and live in houses with inadequate auditory and visual inculation and a closer proximity to neighbors. Persons who felt crowded in a home spent more time in small towns as a child and had lived in their present neighborhood for a shorter period. Those who had spent more time in a city preferred more anonymity and noninvolvement with neighbors than did small town residents. "For this relatively low-density suburban sample, noise of, distance from, and visibility of neighbors were major physical variables related to low perceived privacy within the neighborhood." The relationship with drawing the drapes indicated the importance of window orientation. Within the home, variables of major importance were number of rooms per person and the ability to insulate noisy and quiet activities from each other. The number of rooms that were visually open to each other was not important to perceptions of privacy or crowding.

MATHER, C.E.   162

"Planning Airport Environs," *Build International,* Vol. 6, No. 5, September/October, 1973, pp. 453-461.

KEY WORDS:

Man	Setting	Relational Concepts	Outcomes
Communities contiguous to airports	Sonic environment	Noise	Noise susceptibility

The current surge in air travel has brought about an increased concern regarding effects of aircraft noise in communities proximate to airports. The major conflict between airport functions and community interests has resulted from aircraft flying at low altitudes during take-off and landing operations. The author summarized a number of research projects in terms of the effects which aircraft noise had on buildings, people, and activities in the vicinity of airports, and reviewed planning actions which are useful in the control of aircraft noise.

A 1971 investigation into aircraft noise levels around Los Angeles International Airport by Hurdle, Lane, and Meecham revealed that "the 100 PNdB (perceived noise decibel unit) contour encompassed an area of 10 square miles and a population of approximately 70,000." Furthermore, in residental communities up to 15 miles from the L.A. airport jet noise exceeded average ambient noise levels by approximately 20 dBA. Based on these findings, plus the results of other studies conducted near Heathrow Airport in London, England; the Santa Monica Airport; and the Sydney, Australia Airport, the effects of aircraft noise were found to vary considerably, depending on many factors. For instance, noise appeared to interfere more with falling off to sleep than with sleep itself, and arousal seemed to be dependent on the stage of sleep, on previous sleep deprivation, and on a person's age. In addition, aircraft noise levels peaking between 85 and 88 PNdB's appeared to be the upper limit of acceptability for those engaging in conversation. On the other hand, no clear evidence existed indicating that work activities are disrupted by such noise levels. The recovery of sick persons is generally believed to be retarded by exposure to noise, but adequate experimental evidence did not substantiate these claims. Furthermore, a study by the Australian House of Representatives Select Committee on Aircraft Noise found that property values in some areas are affected by aircraft noise, although each State's Valuer-General stated that aircraft noise was not a criterion used in their valuation of property.

The results from these surveys revealed that some of the concerns over aircraft noise are justified and need to be examined further. By their nature airports are interchange points between different means of transportation, and therefore need to be easily accessible. The author concluded that the most likely solution to community disturbance from aircraft noise is quiet airplanes which can ascend and land steeply.

MATHEWS, K.E. AND L.K. CANON   163

"Environmental Noise Level as a Determinant of Helping Behavior," *Journal of Personality and Social Psychology,* Vo. 32, No. 4, 1975, pp. 571-577.

KEY WORDS:

Man	Setting	Relational Concepts	Outcomes
Physically disabled	Outdoor environ-ment	Noise levels, helping behavior	Interpersonal relationships

Research on the psychological and social effect of ambient noise level was discussed in this paper. The researchers hypothesized that high noise levels would lead to lessened attention to the incidental social cues that structure and guide significant aspects of interpersonal behavior. Conversely, perception of, and responsiveness to central or salient events is not hindered.

The effects of various levels of noise on simple helping behavior were explored in a laboratory and a field setting for a total of 132 subjects. In both experiments subjects exposed to 85 db white noise were less likely than those in lower-noise conditions to offer assistance to a person in need.

With noisy environments, individuals may become less aware of relatively subtle cues produced in interpersonal interactions that more clearly define other's meanings, intentions and behavior. Ongoing behavior and/or interaction would be less flexible and less likely to change to a new direction, since individuals would be less attentive to events that are not directly related to ongoing activities.

Both the laboratory and field experiments confirmed one another. Alternative accounts in terms of the effect of noise on mood and on drive level were also considered.

"Images of the Neighborhood and City Among Black-, Anglo-, and Mexican-American Children," _Environment and Behavior,_ Vol. 3, No. 4, December, 1972.

KEY WORDS:

Man	Setting	Relational Concepts	Outcomes
Children	_Ethnically mixed neighborhood_	_Cognitive maps, perception_	_Map utility_

This study explored children's imagery or "mental pictures" of their world including their homes, neighborhood, trips to school, their city and their favorite and disliked places in the community. The attitudes obtained were explored across age, sex, and ethnic classification.

Ninety-three children from Black-, Anglo-, and Mexican-American families were interviewed. Demographic information including name, date of birth, sex, home residence, school attended last year, number of brothers and sisters, means of transportation to school and whether or not the mother worked outside the home during the day. The children were instructed by the interviewer to draw a map of their neighborhood and of where they live. They were also asked to draw a map of the city of Houston and to describe the trip from their home to school.

From the data collected it was hypothesized that the Anglo children possess a life style and an imagery of their environment which may be considered highly complex in comparison with the Black children. The Mexican-American children defy similar categorization. Most of their performance was intermediate between the Anglos and Blacks, making precise interpretation difficult.

Highly significant ethnic differences occurred in the percentage of the map devoted to the home. Black children drew homes which occupied, on the average, 25% of the neighborhood. Mexican-American children drew homes averaging 5.1% of the neighorhood map and Anglo homes consumed only 2.5% of the map. Significant differences occurred in the propensity of the three ethnic groups to draw a map of Houston. No Mexican-American child and only one Black child attempted this task which was attempted by 24% of the Anglo children.

According to the researchers, the map drawings and descriptions of the trip to school appeared to offer an excellent means of gathering information as to the imagery and perception of the environment as it was viewed by children.

"Light, Radiation, and Dental Caries," _Academic Therapy,_ Vol. 10, No. 4, 1974, pp. 441-448.

KEY WORDS:

Man	Setting	Relational Concepts	Outcomes
Children	Schools	Flourescent lighting	Behavior, Dental caries

A series of studies were carried out to determine the effects of floursecent lighting patterns on the behavior and dental health of elementary school children. It was hypothesized, using the results of past experiments in animal and human behavior that flourescent lighting has negative effects on both behavior and health.

Eighty-seven first graders from four Sarasota, Florida schools were used as the subjects of this study. Two classrooms (control) were lighted with standard cool white flourescent fixtures while the remaining two classes (experiment) were fitted with special Vita-Lite fluorescent bulbs which provided a full spectrum lighting system. Both rooms provided essentially the same footcandle levels and both groups of students were matched statistically in demographic distributions. The study lasted from January to June and observations of hyperactivity were video taped at random periods while academic ratings were collected before and after the test using standardized achievement tests. In addition, the dental health test was continued into a second academic year to monitor the progress of the children's dental caries.

The results tended to support the original hypothesis that standard fluorescent lighting systems were negatively affecting the youths' behavior and dental health. Hyperactivity decreased in the full spectrum of lighted classrooms, when compared to the control unit and academic achievement was shown to be significantly different although not as closely related to the lighting factor as in the hyperactivity measurement. A strong correlation of dental caries to constant exposure to the fluorescent lighting was shown to exist over the full 18 month test period. For those students spending both years in the flourescent setting, a 2.45 cavity/student ratio was found as compared to a .8 cavity/student found in students exposed to the full specturm lighting over the same period.

MAWBY, R.I.   166

"Defensible Space: A Theoretical and Empirical Appraisal," _Urban Studies,_ Vol. 14, No. 2, June, 1977, pp. 169-179.

KEY WORDS:

Man	Setting	Relational Concepts	Outcomes
Victims of crimes/police files	Residential areas, England	Defensible space	Physical layout, importance of high-rise developments, problems of gardens, use of

Data from the second stage of a Sheffield, England survey on area crime patterns served as a basis for this study. Victims of crimes living in nine different residential areas of Sheffield in 1971 were surveyed in an attempt to determine whether area crime patterns varied with design features of the particular neighborhood.

Oscar Newman's book, Defensible Space, asserts that physical layout determines to a great extent where crimes take place. Newman's emphasis was on the problems of high-rise developments. In this study, the high-rises were not altogether at a disadvantage if the development included a shopping district in the middle of the complex, overlooked by the high-rise buildings. Corporate offenses were more likely to be reported in this type of development due to the possibility that a number of people might see any crime and report it. Would-be offenders tended to stay away from such housing complexes.

The differences noted as a result of the survey of victims were; 1) Thefts of and from vehicles were more common in the high-rise areas due to cars being parked some distance from resident's apartment. 2) Thefts from doorsteps/doorways were also more prevalent in the high-rises where there were no gardens to act as a buffer zone. Gardens seem to provide a symbolic barrier to petty theft from doorways. 3) Gardens may also provide a hiding place for offenders who need time and shelter for breaking and entering. 4) High-rise dwellers did not experience as much housebreaking as the low-rise dwellers. 5) "Defensible space" needs more.rigorous definition since many designs for area housing seem to combine both good and bad defensible space qualities.

MACALLISTER, R.J., E.W. BUTLER, AND E.J. KAISER   167

"The Adaptation of Women to Residential Mobility," Journal of Marriage and the Family, Vol. 35, No. 2, May, 1973, pp. 197-204.

KEY WORDS:

Man	Setting	Relational Concepts	Outcomes
Women	Residential households	Mobility and adaptation	Coping behavior/ adjustment

This paper sought to answer the following questions: Is residential mobility disruptive of social relations, and if so, what are the patterns of adaptation to this disturbance? This inquiry was purposely limited to informal social relations as a mechanism of adjustment. The authors provided an overview of relevant literature in the areas of mobility, migration, and environmental change. The bulk of prior research efforts in these areas have possessed built-in assumptions that local moving is nondisruptive, creating little social dislocation, while migration is highly disruptive and forces many social and behavioral changes. The authors disagreed with this assumption, stating that disruption is as likely to occur in local mobility as in international migration and that in fact every change of residence requires some types of adjustment to the new environment. Frequence of interaction with neighbors was also measured.

The sample consisted of nearly 500 persons that represented a
cross-section of married American women. Average age was 42.7 years;
those with children living in the home averaged 2.65. The study was
conducted at two points in time (1966 and 1969). This two-wave national
survey of residential mobility patterns and preferences was conducted
via interviews that obtained a wide array of data. The respondents in
this study participated in both 1966 and 1969 stages of the experiment.
A series of tables reported frequency distributions in both years for
both movers and nonmovers.

It was found that women who moved between 1966 and 1969 were more
frequently sociable both before and after their move than those who did
not move. Also, differential patterns of disruption held for intra-and
extra neighborhood contacts. There was a period of heightened social
interaction on the part of the most recent movers. The authors presented two principal ways to explain this overall pattern of increased
interaction: 1) either these women moved to a new social context that
put them in contact with more compatible neighbors; or 2) the increased
interaction was simply one mode of adjustment to the new environment.
The role of children in this relationship is a factor which operated for
newer residents only. It was concluded that spatial mobility did exert
changes on the social lives of women in these households. It was evident that a heightened search for social contacts on arrival diminished
after an initial time period, which then allowed for development of a
deepened social life in the new residential environment.

MCCARTHY, D. AND S. SAEGERT   168

"Residential Density, Social Overload, and Social Withdrawal," *Human
Ecology,* Vol. 6, No. 3, 1978, pp. 253-272.

KEY WORDS:

Man	Setting	Relational Concepts	Outcomes
Apartment dwellers	*Apartments*	*Density*	*Social overload, withdrawal*

The authors hypothesized that tenants of high rise apartment structures are exposed to more unwanted and uncontrollable social interactions than those of walk-up buildings and will consequently have a
higher preception of crowding, feelings of less control and safety and a
higher degree of social overload. As a result it is suggested in the
literature that the high-rise dwellers will suffer a greater degree of
social withdrawal and alienation than those of the walk-up dwellers.

Sixty interviews were conducted among the 65,980 residents of a
Bronx public housing project. Half of the tenants dwelled in 14-story
high-rise towers while the other half dwelled in three-story walk-up
buildings. Subjects were administered questionnaires which measured
their perceptions of privacy, safety, social and economic control,
social relations, and feelings of environmental satisfaction. Sixty-two
percent of the subjects were black, 19 percent white and 18 percent
Puerto Rican and were all of a similar socio-economic level.

It was found that a high degree of variance existed between the high-rise and walk-up tenants in all areas of interest. High-rise tenants were satisfied with their present dwelling (43.3%), while 93.3% of the walk-up tenants responded similarly. A similar correlation was found in the number of social contacts that the tenants established, with 72.4 percent in the walk-ups and only 48.3 percent in the high-rise. The authors concluded that high-rise environments have a very isolating effect on their tenants and increase their perceptions of crowding and social fear. The data supported this view, especially in the frequency and types of social contacts which were found between the two study groups.

MCKAIN, J.L.   169

"Relocation in the Military: Alienation and Family Problems," *Journal of Marriage and the Family,* Vol. 35, No. 2, May, 1973, pp. 205-209.

KEY WORDS:

Man	Setting	Relational Concepts	Outcomes
Army families	*Residential portions of army*	*Relocation/ alientation/mal- adjustment*	*Alienated wife- mother/community isolation*

The study assessed the relationship between feelings of alienation and family problems associated with moving. A questionnaire was administered (through the mail) to 200 randomly selected army families and a series of 29 interviews were conducted. The problems associated with the military's needs to relocate families and the consequences of these moves were first brought to the author's attention while working with families who were reacting adversely to their assignment in Europe. It was noted that many families, and individual members of families, were isolated from or actually hostile in their relations, (i.e., lack of friendships, environmental disorder, and feelings of rejection, anxiety, and rigidity).

Relevant literature was discussed which focused on spatial mobility and the element of alienation and its deleterious effect on the individual or family. The research questions and hypotheses tested in this study fell within the theoretical framework of anomie. According to the author "anomie theory had its beginning in the work of Emile Durkheim, who attempted to account for incongruencies or deviance in societies on the basis of deregulation or relative normlessness in a social group." The view adopted by the author conceptualizes anomie as a state of mind rather than a state of society.

Correlations were explored between these variables: 1) the wife-mother's feelings of alienation; 2) her lack of military community identification; and 3) family problems, especially those associated with a move. The wife-mothers had a mean age of 25.5, average length of marriages was 5 years, with a mean of 1.8 children. The majority had made six or more moves as an army family. A series of pre-developed and tested checklist and ranking scales comprised the questionnaire. Interview data were collected in an unstructured manner to allow for in-depth open-ended responses.

The computer analysis of the data revealed strong and direct correlations between feelings of alienation, lack of community identification, and family problems associated with moving. Feelings of alienation and lack of community identification, while different, functioned as similar entities in families that just completed a move. Recently occurring personal and marital problems did not correlate in the settled group. Alienated wife-mothers reported higher discrepancy between their needs and satisfactions than low-alienated subjects ($r = .62$, $p<.01$) and more marital role tension ($r = .46$, $p<.01$). In conclusion, the spatial mobility and family problems associated with moving to a new environment were more likely to be found in the Army family where the wife-mother felt estranged from society and the Army community. It is suggested that neighborhood "sponsors" might help these persons ease into new surroundings more comfortably.

MEADE, S.   170

"Medical Geography as Human Ecology: The Dimensions of Population Movement," *Geographical Review,* Vol. 67, No. 4, October, 1977, pp. 379-393.

KEY WORDS:

Man	Setting	Relational Concepts	Outcomes
Malaysians	*Agricultural settlers*	*Migration*	*Morbidity*

Environmental factors such as air quality, site drainage, and transportation systems are causes of human illness. The relationship between man and his environment often determines the degree to which he is able to survive in that environment and what effects will result from his movement among different settings. The article introduced the concepts of man's adaptive characteristics involving environment, population, culture, and time.

Village and land settlement populations were studied in an agricultural area of Malaysia. Four basic microenvironments (home, school, field, and forest) were defined and measurements of illness found in each setting were catalogued. Contagious diseases, for example, were found to exist in places of public gathering and transmitted to home environments by students and shoppers. A set of specific environment related diseases were presented for each subgroup studied.

Mobility was defined as the most significant aspect of the spread and quantity of human disease. Contact between infected and susceptible populations was compared between village and land settlement groups. Although basic human services (i.e., markets) in the village accounted for a major portion of the communicable diseases, land settlers suffered from the effects of long transportation routes (i.e., accidents). The author demonstrated that the communication between different microenvironments can result in human illness through accident as well as through contagion.

The article concluded with a series of recommendations for the planning of human settlements. Environmental causes of diseases and accidents were regarded as predictable and consequences of large-scale population movements can be rationally planned.

MEHRABIAN, ALBERT AND SHIRLEY G. DIAMOND   171

"Seating Arrangement and Conversation," *Sociometry*, Vol. 34, No. 2, June, 1971, pp. 281-89.

KEY WORDS:

Man	Setting	Relational Concepts	Outcomes
University students	*Laboratory*	*Seating proximities; affiliative tendency*	*Preferences*

This study examined seating preferences: 1. as a function of sex; 2. sensitivity to rejection; and 3. the effect of seating choice on conversation. As part of an undergraduate psychology course requirement, 124 male and 120 female University of California students took part in an experiment using a room which permitted observation by way of one-way mirrors. Subjects thought the experiment involved tasks concerning music listening and were told to have a seat in the room until others arrived. Each group was observed for five minutes prior to the playing of music. A questionnaire was administered later which included measures of affiliative tendency and sensitivity to rejection.

Males were found to sit at an average distance of 5.60 feet from others, whereas females sat significantly closer at 5.11 feet. High scorers on affiliative tendency sat at a mean distance of 5.12 while low scorers sat further away at 5.61 feet.

Results would appeared to indicate that individual differences contributed to seating choice, primarily in terms of preference for closeness, which were also affected by sex or affiliative tendency. However, seat selection was not always a free choice, particularly for the third and fourth member of each group of four that was observed. There was more conversation between persons seated closer to each other. Indications were that, in four person groups, females and more affiliative subjects had greater preference for proximity and that immediate positioning of strangers lead them to communicate a liking for each other.

"Intentions and Expectations in Differential Residential Selection,"
Journal of Marriage and the Family, Vol. 35, No. 2, May, 1973, pp.
189-196.

KEY WORDS:

Man	Setting	Relational Concepts	Outcomes
Families	*Urban area/ housing*	*Differential residential selection*	*Intentions; preferences; expectations; actual behaviors*

This paper summarized intentions and expectations in residential
selection among 761 families who had chosen to move (but had not yet
done so) to other areas within or near to metro Toronto, concentrating
on the wives' viewpoint and also comparing it to the husbands' view of
the upcoming move. A background discussion of the act of moving and
before-after aspects and effects was theoretically linked to the field
of human ecology (pre-WW II) and later to the self-selection process of
voluntaristic mobility behavior.

The data reported here came from the first stage of a longitudinal
study of self-selection and subsequent adaptation with respect to physical environments differing in housing type (single-family versus high-
rise apartments) and location (central versus far suburban). The sample
was limited to persons who had greater than average economic choice with
respect to housing location. The distribution of families in the sample
was as follows: 1) High-rise downtown - 109 (14.3 percent); 2) Single
house downtown - 94 (12.4 percent); 3) High-rise suburb - 286 (37.6 percent); and 4) Single house suburb - 272 (35.8 percent). At the time of
the initial phase of the study (1969), median family income was $13,000,
with about 40 percent over $15,000. When interviews were conducted with
a family, separate interviews were held with the wife, husband and one
child (if present in the family). A Master and Orientation Code was
created which served to classify subjects' responses to the sets of
open-ended questions.

The major variables explored and the respective sets of relevant
variables of lesser magnitude were: 1) Change in the type of dwelling
unit as reported as reasons for moving away (size or amount of floor
space, noise and dirt, preference for dwelling type, preference for
tenure, quality of management, type of people in area, and job transfer); 2) Reasons for change of new residence by change in housing type
(size of unit, layout of unit, recreational facilities of building or
home, preference for sector of city, and access to employment); 3)
Change in location (this did not appear to be as important a housing
change in differentiating respondents according to their expectations
and selection process); 4) Expectation of areas of change in weekend
life (home entertaining, going out, outdoor non-athletic leisure, sports
participation, no change, and other); 5) Mobility history and interurban
versus intraurban mobility; and 6) Extent of husband-wife agreement in
general catagories of reasons for choice of a new residence.

Wives appeared to assess alternatives in the selection process
rationally, to be aware of limitations in housing and location they

would experience, and to have expectations about behavioral changes consistent with the degree of change represented by their destination environments. It was found that husband-wife differences suggested the possibility of different methods of coping with post-move adjustment to a new environment. Results also indicated that wives and husbands generally agreed regarding the physical attributes of the actual dwelling, but otherwise possessed different criteria in the housing selection process. The authors concluded that later phases of this study would address needed factors regarding the dynamics of the relationship between pre-move orientations and subsequent behavior and adaptation to housing environments.

MITCHELL, J.K.   173

"Adjustment to New Physical Environments Beyond the Metropolitan
Fringe," *Geographical Review,* Vol. 66, No. 1, January, 1976, pp. 18-31.

KEY WORDS:

Man	Setting	Relational Concepts	Outcomes
Suburban immigrants	*Suburbs*	*Environmental clues*	*Perceptions*

    Areas of suburbia settled by recent city dwellers are usually done quite rapidly and with little preplanning. New communities are developed and filled with occupants on the basis of economic expediancy. This study addresses the problems encountered by these occupants of new exurban communities in terms of how they are able to interpret the surrounding environment and cope with unfamiliar natural hazards.

    A stratified sample of 263 permanent residents in seven Ocean County, New Jersey, communities were questioned concerning their attitudes toward their environment. Ocean County is an area adjacent to New York City and, until the 1940's, was an undeveloped rural backwater. It has grown rapidly since that time; attracting retirees, second homeowners, and families in fairly equal proportions. The retired populations tended to settle in tract housing built in lagoon fills and segregated from other settlements. Most of the developments have been characterized by tract housing methods, resulting in lowered natural vegetation cover and overburdening of public services.

    The author found major attitude differences between long-time residents and recent occupants of the tract developments. "Lagoon dwellers exhibit more conservative recreation site preferences, fail to recognize threatening natural hazards, do not participate in environmental issues, and are more socially isolated." He postulated that occupants of new exurban developments were lured to their environment by unprepossessing attitudes and their socially isolated and apathetic attitudes continue, regardless of surrounding environmental quality. He also concluded that there seemed to be an inverse correlation between social isolation and knowledge about one's physical surroundings. The longer people reside in urban centers, the more knowledge they accummulate about their physical surroundings.

"Some Social Implications of High Density Housing," _American Sociological Review,_ Vol. 36, No. 1, February, 1971, pp. 18-29.

KEY WORDS:

Man	Setting	Relational Concepts	Outcomes
Adults; children	Urban area	Overcrowding	Relationships; effects

The implications that crowding in relation to the microenvironment have for the social and psychological well-being of people have been the subject of much debate. Based on the assertion that "there have been very few well designed studies concerning the personal and family effects attributable to housing in general and the physical features of housing in particular, especially densities," the study examined widely divergent housing conditions within the extremely dense residential setting of Hong Kong.

Information was gathered on the basis of three large-scale sample surveys. The first one consisted of 3,966 people who were at least 18 years of age. The second survey interviewed 561 husband-wife pairs, as well as 2,361 other people who were married and living with their spouse, but whose spouse was not interviewed. The third part consisted of a 10 percent sample of all "Form 3 and 5 pupils in the Colony."

In order to examine emotional strain within the sample studied, three measures were utilized: 1) questions designed to ascertain levels of happiness and worry; 2) information was obtained about standard psychosomatic symptoms; and 3) two separate indices of role or behavioral withdrawal from family and work roles were developed. These measures were then correlated to such variables as the number of square feet in the dwelling unit, the number of people living in the dwelling unit, the number of people per bed, amenities in the room, and the internal partitioning of the dwelling unit.

After various controls were applied, it was determined that densities within dwelling units appeared to have only a slight range of effects. That is, although high density had some effects on the superficial emotional characteristics, such as worry and happiness, high densities did not alter the deeper levels of emotional strain and hostility. Furthermore, little difference was found concerning patterns of husband-wife interaction, but since higher densities resulted in a lowering of surveillence of children by their parents, it was hypothesized that parent-child relationships were altered. In short, the authors concluded that "individuals can apparently tolerate very high densities within their own family dwelling unit, but these high densities may create a street environment that is socially unhealthy to the community by discouraging interaction and friendship practices among neighbors and friends."

MORGAN, J.C., J.S. LOCKARD, C.E. FAHRENBRUCH, AND J.L. SMITH   175

"Hitch-Hiking: Social Signals at a Distance," _Bulletin of Psychonomic Society,_ Vol. 5, No. 6, 1975, pp. 459-61.

KEY WORDS:

Man	Setting	Relational Concepts	Outcomes
Automobile drivers	_Roadside_	_Communication_	_Assistance offered hitch-hikers_

Both short and long range communication is required by humans for survival. Hitch-hiking, an example of long range communication, is comprised of interactions which can be readily observed and quantified. Previous studies and preliminary observations of actual hitch-hikers suggested that some signals are more effective than others in obtaining transportation from strangers.

A pilot study used two experimental designs to assess the importance of sex, eye contact, eating fruit, and driver characteristics in obtaining a ride. Females with eye contact obtained the most rides, males without eye contact obtained the least. There was a significant relationship between eye contact and obtaining a ride in both males and females. Success rates for the two designs were 8.33% and 6.88%. This study suggested the inclusion of other relevant variables (secondary sex characteristics and hand gestures) in the main study.

The main study variables were arranged in a counterbalanced, Graeco-Latin square experimental design. The Graeco components were hand gestures and eye contact. The Latin components consisted of secondary sex characteristics. The hitch-hiker (one of the twelve combinations) stood alone on the road and solicited rides. They and an inconspicuous observer were responsible for collecting data on: the number of cars passing the hitch-hiker, the number of motorists offering rides, observed age and sex of the driver, number and ages of passengers, and the type of vehicle.

Two hundred and eight ride offers were obtained from 4,008 cars. Across all conditions females received approximately three times as many rides as did males. Secondary sex characteristics showed a positive but non-significant trend. A begging hand gesture regardless of condition showed a statistically non-significant inverse trend. Females in the augmented bust, eye contact condition received twice the average number of ride offers, 1 in 10 vs. 1 in 20. Analysis of demographic data indicates the most likely people to pick up hitch-hikers are: 1) male drivers, 2) persons between the ages of 22 and 30, 3) motorists without passengers (esp. children under ten), 4) persons driving 3-4 year old cars, 5) drivers of sedans. Male motorists offered significantly more rides to female vs. male hitchhikers. No trends were apparent for the few female motorists who offered rides.

These findings suggest that the effective signals in hitch-hiking are those that maximize safety and interest and minimize danger.

"Housing Norms, Housing Satisfaction and the Propensity to Move,"
Journal of Marriage and the Family, Vol. 38, No. 2, May, 1976, pp.
309-319.

KEY WORDS:

Man	Setting	Relational Concepts	Outcomes
Families	Metropolitan area/housing	Satisfaction/ propensity to move	Normative housing deficit model

This paper tested a model based on the hypothesis that a family's
dissatisfaction with their housing is the prime determinant of the propensity to engage in residential mobility. Residential mobility was
defined as the extent to which housing failed to meet cultural standards
and referred only to local moves. According to the model that was
tested, there are two criteria used by families to judge their housing:
1) cultural norms, and 2) family norms. Each family constantly evalu-
ates its housing to judge whether it is satisfied with its situation regarding these criteria. If the family's housing failed to meet these
normatively derived needs, a normative housing deficit was said to
exist. The authors discussed three types of behavioral responses when
this occurs: 1) residential mobility (moving), 2) residential adaption,
and 3) family adaptation. Persons' satisfaction with housing norms and
the willingness to move was studied as one of the measures used by families to maintain optimal housing conditions for themselves.

A series of housing variables were presented and supported using
relevant findings in the literature. The major issues and areas of concern that gave rise to the development of the testing model were: 1)
the propensity to move; 2) housing norms and normative housing deficits;
3) normative housing deficits and the propensity to move; and 4) housing
satisfaction as an intervening factor. The model that was developed included five sets of variables. These were: 1) socioeconomic and demographic exogenous variables; 2) intervening variables; 3) intermediate
variables that measure satisfaction and dependent variables; 4) the
desire to move; and 5) moving expectations. Data were gathered from a
sample of households in a metropolitan county outside a major city. The
data base was a two-stage cluster sample of 405 households. The wife of
the head of the household or the female head was interviewed in early
1971. Criteria were developed to determine each family's eligibility to
participate in the study. The key dependent variables were: 1) the
desire to move; and 2) moving expectations. Only 56.3 percent of the
sample wanted to remain in the present neighborhood and the present
dwelling. About 12 percent expected to move within one year and 15 percent expected to move within five years. Intermediate variables were:
1) housing satisfaction, and 2) neighborhood satisfaction. Intervening
variables were: 1) bedroom space deficit; 2) too much living space
(positive structure deficit); 3) too little overall dwelling space (negative structure deficit); 4) owners who wished that they were renters
(owner deficit); 5) renters who wished they were owners (renter deficit); and 6) families who moved during the previous 12 months (recent
movers). Background demographic data served as exogenous variables. A
series of multiple regression correlations analyzed five levels of variables. The normative housing deficit model was the outcome of these
measures.

The findings supported the use of residential satisfaction and normative housing deficits as predictors of the propensity to move. Also,
the propensity to move as a response to housing satisfaction which, in
turn, was a response to significant levels of misfit which occurred between a family's present housing condition, and desired (or optimal)
normative family housing environment as perceived by subjects.

MORRIS, E.W.   177

"Mobility, Fertility and Residential Crowding," *Sociology and Social*
Research, Vol. 61, No. 3, April, 1977, pp. 363-379.

KEY WORDS:

Man	Setting	Relational Concepts	Outcomes
Females	*Rural villages*	*Crowding*	*Fertility behavior/ residential mobil- ity*

This paper analyzed the influence of the lack of space in a dwelling on human fertility and residential mobility. The focus of the study
was on links between: 1) adaptation behavior of the family, 2) control
of family size, and 3) mobility of the family unit.

The data were based on 395 interviews (45-75 min. in length) of
female heads of households or wives who came from approximately 10% of
the households in the small villages of Tioga County, New York, in 1971.
No woman over 64 was interviewed. Dependent variables were births and
actual moves made. Independent variables included such factors as husband's education, occupation of head of household, single-family dwellings, owners, and bedroom deficit.

Regression analyses predicting fertility and mobility indicated
that the ability to move from one residence to another serves to relieve
the negative type of pressure that residential crowding might exert on
fertility behavior. If residential mobility were to be severely hampered by a shortage of housing of the sizes and price ranges needed, it is
possible that fertility might be negatively affected by crowded conditions in these homes. Therefore, present findings are partially due to
historical facts of the time which include a range of houses of different sizes available and a preference of families in the area to move to
larger quarters rather than curtail family size.

The results of this present study indicated that under conditions
of moderately plentiful housing, great freedom of movement, and moderate
levels of crowding, the lack of space does not seem to affect fertility
rates.

"Death Attitudes of Residential and Non-Residential Rural Aged Persons,"
Psychological Reports, Vol. 43, No. 3, December, 1978, pp. 1235-1238.

KEY WORDS:

Man	Setting	Relational Concepts	Outcomes
Elderly	*Rural institu-tional/non-insti-tutional*	*Death attitudes*	*No differentiation*

Questionnaires concerning death and death anxiety were administered
to 40 volunteer ambulatory institutionalized and 40 ambulatory non-
institutionalized aged persons. Residents in group one lived in a
Wyoming State facility for the aged, while group two members belonged to
a community senior citizens center. On each measure, the two groups
could not be differentiated. Each subject was administered a revised
version of Middletons questionnaire to which seven items were added and
Templers Death Anxiety Scale. Subjects generally reported a realistic
view of death with no pronounced fear of or preoccupation with the subject. Only 5% responded that they frequently wished for death, while
82% thought about living things and looked forward to the future. A
series of items was reported with results associated with each particular predictor of death attitudes.

Of all respondents, only 20% indicated that they were nervous or
anxious when others talk about death. A number of variables focusing
on mobility behavior pattern, and activities in both environmental settings were studied. Results indicated that within this sample of institutionalized and non-institutionalized rural aged who reported themselves in fairly good physical health, no differences were detected in
attitudes toward death associated with mode of residence. It was concluded that influences of the physical environment were not pronounced
in this study of death attitudes of the elderly.

"Recommendations for the Admission and Control of Sunlight in Buildings," *Building and Environment,* Vol. 11, No. 2, 1976, pp. 91-101.

KEY WORDS:

Man	Setting	Relational Concepts	Outcomes
Human	*Buildings*	*Sunlight exposure*	*Visual preference; design revision*

In 1945 the British Standard Code of Practice on Sunlighting was
established in order to form the basis for allowing desirable amounts of
sunlight into various types of structures. Since that time there have
been many changes in the types of buildings erected, as well as changes
in attitudes regarding environmental standards. Citing the fact that

many recent studies have concentrated on the thermal aspects of sunshine, the authors examined the visual aspects of sun penetration for four building types: hospitals, schools, offices, and housing.

The three year study consisted of three simultaneous research components: 1) field work was undertaken in existing buildings by means of a social survey, and an assessment team which measured the visual and thermal environment; 2) laboratory experiments which enabled visual and thermal measurements to be taken of a full-sized control environment; and 3) computational analysis of sunshine availability, and the cataloguing of existing sunlight control devices.

Results indicated that the desirable amount of sunlight will vary according to activity groups. It was found that the more confined the activity, the more severe the adverse effects of sunshine may be. Furthermore, by analyzing various factors it was determined that "a figure between 1/4 and 1/3 of the maximum yearly average number of available sunshine hours can be the best compromise between desire for sunshine, need for control, and design practicality. The figures of 400 to 500 hours therefore have been taken as the recommended minimum specified number of admitted sunshine." Consequently, since windows facing northeast and northwest cannot admit more than 300 sunny hours, certain ranges of orientations would not be in compliance with the recommendations of this study. Realizing that planning for sunlight in buildings has to be closely tied to other aspects of design, the authors conclude that, "it is not intended that the recommendations should restrict the scope of possibilities in sunlight design too rigorously but rather should assist the designer to avoid problems or to appreciate them should other requirements take precedence."

NE'EMAN, E., J. CRADDOCK, AND R.G. HOPKINSON   180

"Sunlight Requirements in Buildings - I.: Social Survey," *Building and Environment*, Vol. 11, No. 4, 1976, pp. 217-238.

KEY WORDS:

Man	Setting	Relational Concepts	Outcomes
Office workers; general population	*Residence, school, office, hospital*	*Sunlight penetration*	*Visual sunlight*

In this article the authors explain in detail the third component of a three-year study in England concerning the visual aspects of sun penetration for various building types (Ne'eman, Light, and Hopkinson; 1976, 91-101). Four separate questionnaires were designed for each of the four settings: housing, schools, offices, and hospitals. The surveys were developed to seek answers to the following questions: "Do people while staying indoors like the sun to shine into buildings they occupy; or on the outdoor scene; what affects their liking; how do they rank the sunlight amenity in their order of preference for environmental satisfaction; and if they do desire the sun to shine indoors, is it for its warmth, light, health, or other benefits?"

A review of prior literature did not reveal much information regarding sunlight penetration preferences. Thus, a preliminary depth interview was implemented consisting of open-ended questions, which allowed the respondents complete freedom to express their feelings and preferences. The responses which were gathered then formed the basis for the four structured questionnaires.

The housing questionnaire was conducted at three different locations, which were chosen to represent a wide range of incomes. Most of the units examined had exposure to sunlight for most of the day, with 50% facing W or SW, 27% facing E or SE, and 18% facing due S. A total of 211 interviews were conducted.

Schools were chosen with the objective of covering a representative range of different geographical and climatic environments. Nine schools were visted, and interviews were carried out in six of them. Information was gathered from a total of 89 teachers.

The office survey utilized 20 bipolar scales. Questionnaires were administered in three large buildings to a total of 263 office workers. Sites were chosen representing a wide range of environmental conditions so that comparisons of window dimensions could be obtained.

Three contrasting wards comprised the hospital study. One was built in the early years of this century and allowed very little sunlight penetration. The second was a pre-World War II design, and although there was some limited sunlight penetration, the ward was a large space containing numerous rows of beds. The third was a modern hospital and was therefore designed with the latest thoughts on hospital design. Fifty-five patients and twenty-nine staff members comprised the study.

In general, results indicated that sunshine promotes psychological well-being, but that it is not a first priority in producing environmental satisfaction when compared with such factors as heating, ventilation, noise, and acceptable views. Nevertheless, the study did emphasize that sunlight should be provided in interiors, with careful consideration given to its possible effects. Furthermore, it was found that the function of a space determines the flexibility of its occupants to adapt to environmental conditions. Consequently, the authors suggest that "any methods designed to provide interiors with sun should consider the activities and work positions existing in the building and avoid causing visual and/or thermal discomfort to the occupants employed in them."

NE'EMAN, E.   181

"Sunlight Requirements in Buildings - II.: Visits of an Assessment Team and Experiments in a Controlled Room," _Building and Environment_, Vol. 12, No. 3, 1977, pp. 147-157.

KEY WORDS:

Man	Setting	Relational Concepts	Outcomes
General population	Visual, school, hospitals, universities	Sun penetration, thermal	Visual preference, thermal comfort

In this article the author explains two components of a three-year study in England concerning the visual aspects of sun penetration for various building types. One part of the research consisted of a team of experienced observers and technical staff who carried out thermal and lighting measurements in a representative selection of buildings. The other research component reported here took place in a controlled laboratory setting which paralleled the on-site experiments so that environmental conditions could be molded to fit all of the desired situations.

The assessment team consisted of from 8-12 observers in each case. This team visited two hospitals, one school, and three universities. During the visits a total of 278 questionnaires regarding measurements of the visual and thermal environments were taken. The visits could only be made on sunny days, but the days selected covered all seasons of the year. Four major conclusions resulted from these measurements: 1) A firm relationship between subjective responses to sunlight and measured luminances on contrasts inside buildings could not be established. 2) Overall comfort of the environment is equally dependent on visual and thermal comfort. This means that both visual and thermal comfort conditions should be satisfied, before overall comfort in the environment can be achieved. 3) Glare from the visual task seems to be the most significant response for the evaluation of visual comfort. 4) The further the bright surfaces in sun are from the observer, the less is their effect on glare discomfort.

Three sets of experiments were undertaken in the controlled setting. For one week, hourly measurements of temperatures were made, and occupants were asked to give an assessed rate of their comfort on the Bedford Comfort Scale.

In the second set of experiments the amount of sunlight which was allowed to enter the room was controlled. In addition, uncomfortable thermal conditions (near 30° C) were produced by means of convector heaters. Each subject in the room made an assessment of visual discomfort and thermal comfort on scales provided at each seating position. The main variables examined included the following: 1) amount of sun (2 levels); 2) temperature (2 levels); 3) seating position (3 levels); 4) glare (2 levels); and 5) subjects (10 levels).

The third set of experiments concentrated on the assessment of visual comfort, with the thermal environment being kept within the human comfort zone. Two levels of glare were also examined.

In general, the results showed that thermal comfort or discomfort did not affect the visual assessments and that there was no significant difference between male and female assessments. There were also indications that the quantity of sunlight penetration into a room is the key variable for both visual and thermal comfort assessments in the space.

"Crowding: Effects of Group Size, Room Size, or Density?" *Journal of Applied Social Psychology*, Vol. 6, No. 2, 1976, pp. 105, 125.

KEY WORDS:

Man	Setting	Relational Concepts	Outcomes
Human	*Rooms*	*Crowding*	*Perception*

This study attempted to identify relationships between room size and density with the perception of crowding and task performance. Using the models of past researchers, the author has set about to test a series of room and crowd size situations among both male and female subjects and measure the results with past hypotheses concerning feelings of crowding.

Five-hundred and ninety-four male and female subjects were divided into groups of fours and tens and exposed to room sizes of 16, 40, and 70 square feet spaces. Each group was seated at a chair and presented with tasks that were to be performed both individually and as a group during the course of the experiment. An attitudinal questionnaire was also administered after each task session to measure the subjects' perceptions of crowding, fear, anger, well-being, and satisfaction.

In general, the study showed that groups which performed the task as a unit in the four member team were much more satisfied in all situations than any other arrangement. Larger group sizes tended to express more feelings of hostility and anger than the four member group. The most negative environment was found to be the large group working individually in a small and densely populated space. Density in itself, however, did not influence a majority of the responses to crowding. Females were more density oriented and preferred a density figure of 10 square feet per person. Males seemed to be more adversely affected by group size and room density. Surprisingly, no significant measures of performance could be attibuted to density. The author suggested that feelings of crowding might be tuned out by the individual when presented with a specific task and the concept of crowding may in fact be a more long term behavioral variable that cannot be measured by short term performance experiments.

NORMAN, K.L.   183

"Attributes in Bus Transportation: Importance Depends on Trip Purpose," *Journal of Applied Psychology*, Vol. 62, No. 2, April, 1977, pg. 164-170.

KEY WORDS:

Man	Setting	Relational Concepts	Outcomes
Students	*Room*	*Bus system effectiveness*	*Attribute judgement model*

It is assumed that the goal of research in this area is to assist the planner by providing relevant information on the consequences of policy decisions. Transit authorities have been concerned with the factors of mode choice in order to predict ridership for bus systems. It is also assumed that three issues must be understood: First, the planner must know what attributes of the system are relevant to the potential patronage and the utilities or subjective values of different levels along these attributes; secondly, the planner must know how individuals integrate this information; thirdly, the planner needs to know how situational variables (trip purpose, etc.) affect the many attributes in determining ridership.

Twenty-four college students judged the probability that they would ride hypothetical bus systems for both a work trip and for a leisure trip. Bus systems varied regarding fare, total walking distance to and from the bus stop, number of stops enroute, and time of service during the day. Sixty bus systems were rated. A questionnaire was used that included a partial rank order step and scaling tasks. The entire subject group rated the systems during the same 30-45 minute session.

An information-integration model was used in which the subjects' responses were assumed to be a weighted geometric average of the subjective values of the attributes describing a given system. The effect of trip purpose was accounted for by changes in the weight and subjective value of information. Results indicated that evidence warranted argument against the methods used in prior studies that focused on similar issues. The approach used here was demonstrated to increase the generality of results by using psychologically meaningful parameters to specify how the _effect_ of an attribute _varies_ from one situation to another.

The authors concluded that although these results have a number of practical implications for bus transportation systems, caution is necessary in generalizing these results to applied problems. Although the same judgement model may apply across two different trip purposes, some judgemental parameters may vary. Consequently, additional research is called for to determine which attributes are situation dependent and how their weights and subjective scales in the judgement model are affected by the contextual situation.

O'LEARY, K.D. AND A. ROSENBAUM   184

"Flourescent Lighting: A Purported Source of Hyperactive Behavior," _Journal of Abnormal Child Psychology_, Vol. 6, No. 3, 1978, pp. 285-289.

KEY WORDS:

Man	Setting	Relational Concepts	Outcomes
Children	Classroom	Flourescent lighting	Hyperactive behavior

A number of studies have purported to show that flourescent lighting systems in the classroom cause an increase in hyperactivity among children with conduct disorders. The authors of this article find that such studies have lacked a rigorous scientific base and purpose. Thus, it was felt that these studies provided insubstantial results relative to this hypothesis.

Seven first-grade children with conduct-related disorders and hyperactivity problems were tested over an eight-week period in a university laboratory school classroom. The lighting conditions varied during the experiment between standard cool-white flourescent fixtures on odd weeks to a daylight simulating flourescent system on even weeks. Behavior of the children was measured by two observers during the test on the task orientation of each child on the activity level present during random periods throughout the day. Reliability checks averaged 82 and 92 percent for each test. A third test consisted of physical measurement at the end of each week of each child's Critical Flicker Fusion (CFF) or that frequency of flicker which an oscillating light is perceived by a subject as a continuous source.

A three-way analysis of variance for the two behavioral measures failed to show a significant difference in the activity of the children between the two lighting sources. The effect of the CFF test did indicate "that children fused at lower rates across weeks in the broad spectrum condition, whereas the children in the standard flourescent condition fused at higher frequencies across weeks." The authors concluded that no significant behavioral distinctions could be drawn from the use of flourescent lighting systems, but theorized that the broad-spectrum lighting was associated with sensory visual fatigue as evidenced by the results of the flicker fusion test.

OLOKO, O.   185

"Influences of Unplanned Versus Planned Factory Locations on Worker Commitment in Nigeria," *Eskistics,* Vol. 36, No. 214, September, 1973, pp. 175-181.

KEY WORDS:

Man	Setting	Relational Concepts	Outcomes
Factory workers	Nigeria	Environmental setting	Worker committment

It has been the practice of many developing countries to concentrate the bulk of their industrial resources in new and highly centralized industrial estates, many times far removed from the population centers in which the workers live. The author of this study hypothesizes that: 1) these isolated and socially remote estates are debilitating to the workers commitment; and 2) a much higher degree of satisfaction will be found in less developed and more socially integrated industrial environments.

Five hundred twenty Nigerian workers from three distinct industrial developments were interviewed concerning their commitment to remaining in and being satisfied with factory work. This index of worker satisfaction (CFW index) measured the workers' preferences of work environments and work situations, satisfaction with their jobs, and their determination to remain in their present work situation. The survey also measured demographic indices such as income, neighborhood, and trade union associations. It was expected that a higher degree of worker satisfaction would be evident from the sample taken at a factory located in the immediate vicinity of a residential area, rather than in either centralized industrial estate or a remote and isolated estate.

The study showed that the residential centered environment had a significantly higher standard of worker commitment (40.7%) than either the isolated or highly planned central estates (28.8% and 21.0%). It was also shown that income was not a significant influence among workers at the low income level nor was the influence of trade unions or neighborhood association relevant to a worker's satisfaction in terms of environmental location. The author concluded that the location and type of industrial estate is a major factor in determining worker satisfaction and commitment over a long period of time.

OLSEN, H. AND R. CARTER   186

"Social Psychological Impact of Geographic Location Among Disadvantaged Rural and Urban Intermediate Grade Children," *Child Study Journal,* Vol. 4, No. 2, 1974, pp. 81-92.

KEY WORDS:

Man	Setting	Relational Concepts	Outcomes
Children	*Urban/rural*	*Location; learned ability*	*Grade level-setting influences learning*

Previous studies have indicated that there exists significant differences between the learning attitudes of children reared in rural or urban environments. The authors of this study argued that in one specific self-perception test (academic ability) urban children will have a higher degree of self-awareness and confidence than those children who come from a more rural background.

One hundred eighty-four intermediate grade school children were subjects for the study. One hundred twenty-five were fourth, fifth, and sixth graders from an economically deprived area of southeastern Ohio, while 59 were fourth, fifth, and sixth grade children from an urban school district in western New York. The Michigan State Self-Concept-of-Academic Ability Scale (SCOAA) was administered to the subjects during June and July of 1971 in order to measure their self-perception of learning and attitudes of achievement. Hypotheses were tested by utilizing a three-way univariate analysis of variance, comparison levels being grade, gender, and geographical background. National statistics have shown that a raw score of 40 on the SCOAA indicates a "superior" attitude, while a score of 20 indicates an "average" rating.

Two comparison levels (grade and location) were shown to be statistically significant ($p < .05$) as factors in determining self-perceptions of academic ability. It was shown that, in support of the original hypothesis, rural children did perform more poorly on the SCOAA than did the urban subjects (mean score of 29.3 versus 30.6). Children

of both locations in the fourth and fifth grades were also more confi-
dent in the test than the sixth graders. The authors concluded that the
results of this study confirm and support Frost's (1971) contention that
rural children perceive their school and educational environment as
limited and are aware that their opportunity for academic success is
restricted.

OLSHAVSKY, R.W., D.B. MACKAY, AND G. SENTELL   187

"Perceptual Maps of Supermarket Locations," _Journal of Applied Psychology,_ Vol. 60, No. 1, February, 1975, pp. 80-86.

KEY WORDS:

Man	Setting	Relational Concepts	Outcomes
Consumers	Urban area	Perceptual/actual distance	Compared perceptions

        Issues in measuring consumers' perceptions of where retail firms
are located were discussed. The authors reviewed past studies that have
speculated on the importance of subjective distance and perceptual maps
in determining shopping behavior as well as findings that support the
hypothesis that perceived locations may differ substantially from actual
locations. In this study, the attributes of the multi-dimensional scaling techniques led to its use as the way of operationally defining subjects' perceptual maps. Two algorithms were used: KYST (Kruskal,
Young, and Seery) and INDSCAL (Carrol and Chang). The principal hypothesis are: 1) that although the maps are expected to be internally consistent, they are expected to differ in content; 2) that the character-
ization of patterns of distortion is possible, and 3) that the time and
energy required to overcome distance from the store usually results in a
negative correlation between the frequency of shopping at stores and the
distance of the store from the consumers residence, and the result is
that this relationship is more pronounced when perceptual distances are
used than when actual distances are used.

        The sample consisted of 55 consumers who were interviewed in Bloom-
ington, Indiana. Individual and aggregate perceptual configurations
were contrasted to an actual map of supermarket locations. Values for
the measures of fit between the actual and aggregate maps under KYST and
INDSCAL resulted in correlation measures that were significant at the
.001 level. This measurement of actual and perceived distances to each
of the six supermarkets signficantly correlated with the frequency of
shopping at the respective stores for each consumer.

        Both common and idiosyncratic types of distortion were apparent
regarding the east-west axis of both scaling measures as well as along
the north-south axis. For 36 of the 55 subjects, correlations between
perceptual distance and frequency of shopping were more significant than
those between actual distance and shopping frequency. This finding
lends support to earlier speculations (particularly in marketing and
geography). The authors concluded that the extent to which behavior is
a function of perceived distance or vice versa remains an issue for further study.

OSTBERG, O. AND A.G. MCNICHOLL   188

"The Preferred Thermal Conditions for 'Morning' and 'Evening' Types of
Subjects During Day and Night - Preliminary Results," _Build Interna-
tional_, Vol. 6, No. 1, January/February, 1973, pp. 147-156.

KEY WORDS:

Man	Setting	Relational Concepts	Outcomes
College students	Controlled thermal	Thermal factors; circadian rhythm	Thermal preference variation

An individual's thermal experience is a function of that person's
heat production and heat loss rate. Consequently, the circadian rhythms
in man's behavior and physiology are thought to be important variables
in the thermoregulation of the human body. It is hypothesized that the
interaction of the influence of food intake and muscular work (heat production) with separate oscillations of heat loss will ultimately determine the level of thermal comfort experienced by an individual. Based
on the notion that man's preferred thermal conditions are dependent upon
body temperature, sixteen subjects were studied individually for two
different times of the day in a climate chamber where the effective temperature could be changed according to their preferences.

Although previous research by Fanger (1973) led him to recommend
that the thermal environment in buildings be held constant throughout
the day, the authors believed that two major reasons could account for
these findings. First, "no studies have been directed to the differences between day and night, and just to look for differences between a.m.
and p.m. is not enough for psychological functions." Secondly, "the
differences between individual subjects have been regarded as error
variances in most experimental results, instead of being treated as useful parameters of variation."

The eight male and eight female subjects who took part in the
experiments were carefully chosen from an initial pool of 110 students
by means of a questionnaire designed to determine lifestyles and activity habits. Four male and female "evening" types were chosen, as were
four male and female "morning" types.

Each of the subjects were given an oral thermometer and were
instructed to take his temperature every 1 1/2 hours during a 48 hour
period so that 32 measurements were gathered in all. One by one the subjects spent two sedentary three-hour periods in a climate chamber, during which time they gave their thermal preferences based on a seven-
point scale. Skin and rectal temperatures were monitored continuously,
weight-loss measured every hour, and the metabolic rate recorded at the
start and end of the test period. The subjects wore standard 0.6 clo
uniforms. The times at which the subjects were studied in the climate
chamber were chosen to coincide with the maxima and minima of their body
temperatures, as determined by the pre-test temperature measures.

Results indicated that the methods used to select subjects and the
time of day for the testing were successful, in that "high/low body temperature was a very significant source of variance in the measured rectal temperature, whereas the morning/evening interactions were not."

Furthermore, the decrease in mean skin temperature was significant despite the very small absolute change (0.04° C). Citing the need for the
testing of a larger number of subjects in this regard the author concluded that "the lower the sedentary body temperature (within the comfort range) the lower the preferred temperature." Furthermore, "it is
thought that a replication of this experiment, with a larger number of
subjects, would show a significant decrease in preferred temperature
during the night."

PAGE, R.A.   189

"Noise and Helping Behavior," *Environment and Behavior,* Vol. 9, No. 3,
September, 1977, pp. 311-334.

KEY WORDS:

Man	Setting	Relational Concepts	Outcomes
Human	Street	Noise	Helping behavior

Noise has often been viewed in prior studies as a factor in the
behavior of people in helping situations. The author of this study
hypothesizes that helping behavior will decrease as noise levels increase
and that males and females will react differently in various noise
situations.

Three experiments were performed to test the above hypothesis. The
first situation involved the introduction of three different noise
levels (50, 80, and 110 dBA) into a pedestrian tunnel heavily travelled
by university students. A female experimenter dropped an object in the
path of 60 male and female subjects over the above noise range and
observations were made of the amount and type of help offered. In the
second experiment, a similar helping situation was created on a city
sidewalk in the presence of 92 dBA construction noise and again during a
72 dBA ambient level. One hundred males and 100 females were again
observed concerning their helping behavior to a female experimenter.
The final experiment consisted of both male and female experimenters
requesting assistance from 200 subjects (male and female) at the same
sidewalk and at the two noise levels.

In conformance to past studies, it was found that males tended to
give physical help in most situations while females assisted with verbal
responses. A significant decrease in female assistance was found in
the high noise conditions (76%) as compared to low noise environments
(92%). Overall, the study showed that as noise increased, the degree
and amount of helping behavior decreased, especially among females in
verbal assistance. The author concluded that to a significant degree,
noise interrupted these communication patterns and was primarily responsible for a physical interruption of social behavior. While showing a
general trend to support the hypothesis, results of the three tests were
not extremely conclusive when summarized.

PALM, R. 190

"Real Estate Agents and Geographical Information," *Geographical Review,*
Vol. 66, No. 3, July, 1976, pp. 266-280.

KEY WORDS:

Man	Setting	Relational Concepts	Outcomes
Real estate agents	Neighborhoods	Perceptions	Preferences

Many people moving to a new environment depend upon real estate
agents as primary sources of information concerning housing types and
neighorhood structure. This study proposed the thesis that real estate
agents of a large and diversified agency are biased in their assessment
of neighborhoods and have a narrow perception of the environment as a
whole.

Two-hundred-fifty realtors from Minneapolis, Minnesota, were interviewed and asked to select living environments for eight hypothetical
clients. The clients ranged in socio-economic rank from a dentist to a
deliveryman. The author was interested in answering three basic
questions concerning the agents' views of the urban environment and how
they matched these perceptions with prospective buyers. The first
question dealt with the range of an agent's perception and the determination of how large a piece of the urban area he was able to
comprehend. Secondly, the study sought to relate neighborhood quality
with an agent's perceptions of that environment's attractiveness to
potential clients. Finally, the study correlated different agent's
views of the same neighborhood type to determine the reasons for local
preferences.

The author concluded that real estate agents, even in a city as
homogeneous as Minneapolis, cover only a small portion of the urban
housing market. Agents tend to evaluate a neighborhood's attractiveness
or social desirability based upon vacancy patterns rather than overall
urban conditions. The agents had a relatively good idea of housing
types available only if the housing stocks were on the market at that
particular point in time. Finally, it was found that real estate agents
tended to evaluate housing stocks based upon specific geographic
location. Most of the agents tended to bias their evaluations in favor
of territories with which they were most familiar and which were closest
to their own offices.

PAMPEL, F.C. AND H. M. CHALDIN 191

"Urban Location and Segregation of the Aged," *Social Forces,* Vol. 56,
No. 4, June, 1978, pp. 1121-1139.

KEY WORDS:

Man	Setting	Relational Concepts	Outcomes
Elderly, urban resi-dents	Urban areas	Segregation/inte-gration	Ecological models

This study examined intraurban location and segregation of the aged in the cities of Cleveland and San Diego using data from the 1970 census for all the residential blocks in both cities. Older persons were predicted to: 1) live near the city centers where they are closer to shopping areas and transportation; 2) in apartment units with little space; 3) in lower value housing; and 4) in densely populated areas near the city centers. This study addresses three questions concerning age-segregation of the elderly. Where are the aged located in urban areas? Are they segregated from other age groups? Does the location and segregation of the aged differ in old and newer cities?

Findings included: 1) Although blocks near the city center have a higher proportion of old people than the peripheral blocks, the differences are not large; 2) In fact, the aging population are relatively dispersed throughout both cities; 3) Old people are moderately segregated from young people no matter where they are located in the cities studies; 4) In San Diego, multi-unit structures and housing value were more centralized and segregated, and the population potential was not related to multi-unit structures and crowding; and 5) In general, both cities were very similar.

The implication of the findings is that previous conceptions of the aged as highly segregated in inner-city ghettos tend to exaggerate the situation. Use of an ecological model to predict location of the older population is an improvement over former methods of study. If more attention is paid to city differences as well as to data on neighborhood populations, the explanatory ability of the ecological theory can be tested.

PAULUS, P., V. COX, G. MCCAIN, AND J. CHANDLER   192

"Some Effects of Crowding in a Prison Environment," *Journal of Applied Social Psychology,* Vol. 5, No. 1, 1975, pp. 86-91.

KEY WORDS:

Man	Setting	Relational Concepts	Outcomes
Prison inmates	*Single cells; dormitories*	*Crowding*	*Negative consequences*

The authors of this article proposed that a sustained amount of crowding with a confined population will result in a desire to expand a subject's immediate environment and at the same time increase the subject's tolerance of crowded spaces. The long term effects of confinement in prison environments was the basis for experimentation concerning this thesis.

One-hundred-forty-two male inmate volunteers at a Federal correction center at Texarkana, Texas, were the subjects of this test. Most men were housed in either a single cell environment (N =48) or in a common dormitory space containing 26 prisoners (N = 73). The subjects were tested three times over a period of two years by means of questionnaires and the manipulation of scaled spatial models. The model test was a representation of the dormitory space which each prisoner experiences during his first 30 days in confinement. The prisoners were asked to place scaled models of the human figure in the room until they thought the space was becoming crowded. Single celled subjects were used to measure the affects of spatial density, while dormitory inhabitants were used as a measure of social density crowding.

The results of the experiments confirmed the original hypothesis that the immediate effects of crowding on the prisoners were negative feelings concerning their environment and a desire to produce a living space with more physical area. The tests seemed to suggest that the inmates in the dorm arrangement experienced negative feelings as a function of their limited time in confinement as well as the highly dense living conditions. The authors suggested that the more tolerant reactions of the single-celled dwellers were a reflection of their adaptive nature as much as of the physical environment.

PEDERSON, D.M.   193

"Effects of Size of Home Town on Environmental Preference," *Perception and Motor Skills,* Vol. 45, 1977, pp. 955-966.

KEY WORDS:

Man	Setting	Relational Concepts	Outcomes
University students	*Laboratory*	*Size of home town*	*Perception of environment*

Past studies have shown that environmental perception is shaped by the experiences that one has with the environment. This perception, in turn, influences one's responses to environments. This study focused on the effects of the size of home town to ratings of five distinct environments.

Three equal groups of 56 college students were selected from volunteers. Subjects differed in size of home town and were classified in the following manner: Small - 0 to 19,999; medium - 20,000 to 99,999; and large -100,000 plus. A 23 item, seven point semantic differential scale, used to rate forest, beach, small town, desert, and large city, was administered to the subjects in small groups. The order in which the environments were ranked was randomly assigned. Twenty-three 3 x 56 x 5 factorial analyses of variance were utilized where factors were size of home town, subjects within size of home town, and repeated measures on the five types of environments. The Newman-Keuls test, which measured the differences between means, was used to test main and interaction effects.

Only one of these 23 F ratios showed significant results for the size of home town main effect. Subjects from large and medium sized cities rated the environments higher overall on the good - bad scale ($p <= .05$). All F's were significant for type of environment since the environments that were rated were heterogeneous. For the 23 scales combined, the order and means of the environments were as follows: Forest (5.94), Beach (5.30), Small town (4.88), Desert (4.65), and Large city (3.97). This order varied depending upon the particular scale. Interaction of size of home town by type of environment was examined in two phases: A) differences among mean ratings across size of home town and B) differences in patterns of mean ratings across environment from one size of home town to another. Across all environments, only the good-bad scale produced a significant size of home town effect. On half of the scales, subjects from all three sizes of home town rated the environments approximately the same as one another. For the other half there was no consensus among all subjects. Subjects from any two sizes of home towns did not consistently view environments in the same way. Many patterns of preference emerged from this analysis.

PERETTI, P.O.   194

"Effects of Controlled Illumination Levels upon Verbal Responses Latency in a Color-Word Interference Task," *Journal of Psychology,* Vol. 95, March, 1977, pp. 189-193.

KEY WORDS:

Man	Setting	Relational Concepts	Outcomes
Elementary students	Laboratory	Illumination	Verbal response

This study attempted to correlate the ambient illumination level with verbal response among grade-school age children. The hypothesis of the study was that extremely high and low lighting levels will interfere with visual performance and that a universally acccepted illumination level will provide a significantly better response environment.

Forty sixth graders from a Chicago school were used as the experimental population and were randomly divided into four groups of 10 each. The first group was used as a pilot or calibration study group, while the three remaining populations were administered a standardized visual response test in 5, 50, and 150 footcandles (fc) of light. The test consisted of verbally identifying numbers and colors within specific time periods. The 50fc group was considered the control group since this level of lighting was documented in past studies as the optimal environment for reading and color identification.

As was predicted, both the low and high illumination levels had significantly ($p < .01$) higher response times (91.5 and 85.2 seconds) when compared to the group performing at the 50fc level (73.7 seconds). A greater number of errors were also found in groups 1 and 2 (9.6 and 7.1) as opposed to those in the control population (4.6). The author concluded that both high and low levels of lighting have adverse effects on hue and saturational qualities of color and that performance on a word-color interference task is probably a function of the ambient light level.

"Do Freeways Make Good Neighbors,"   *Traffic Engineering,* Vol. 43, No. 9, June, 1973, pp. 20-23.

KEY WORDS:

Man	Setting	Relational Concepts	Outcomes
Residents	*Freeway environs*	*Proximate/ attitudes*	*Self-image*

A random sampling was made of three different types of neighborhoods which bordered on freeways in a southwest metropolitan community. Personal interviewing of either head of household or spouse was accomplished in 258 households.

In the three neighorhoods, no one lived more that 799 feet from a freeway and the mean was 300-400 feet from the freeway. Neighborhood one was predominantly lower social class Mexican-Americans living next to an elevated-fill/above-grade freeway. Neighborhood two was all black families living next to a depressed or below-grade highway. The third neighborhood was all-white middle or upper lower class, most of whom moved to the area after the freeway was built.

The participants were shown simple line drawings of four different freeway designs which were called: (1) above grade (elevated fill), (2) below grade, (3) on-structure (elevated), and (4) at grade.

They were asked to state their preference for a type of highway as regards living in close proximity. A computer analysis was performed on the data and results were as follows: (1) 40% living in the third (at grade) area preferred living "at-grade", (2) 32% in the second area (depressed/below grade) preferred living there to living near other types of freeways, (3) 30% of those in the first area (elevated-fill/ above-grade) preferred that area but 30% in that same neighborhood preferred another kind of freeway to live near, (4) the elevated-structure/ above-grade freeway was least preferred of all four types - even when none of the participants lived near such a structure, and (5) reasons for preference or dislike: Depressed/below-grade freeway was the quietest but the dirtiest and worst for conservation and at-grade freeways were rated as safest by subjects. Elevated freeways of any kind were the most dangerous and caused the greatest neighborhood deterioration but were best for conservation.

Overall results of the survey indicated that two-thirds of the persons interviewed in all three neighborhoods felt positive about the presence of the freeways and also felt that the greatest impact of the freeways was on property values which were perceived as enhanced by the proximity of easy access to a major highway. Persons living near the at-grade freeway saw more advantages than the others. It was concluded that planners of communities near freeways need to give accurate information to residents since the factors involving living near a freeway are not beneficial to all.

PETERSON, G.L., BISHOP, R.L., MICHAELS, R.M. AND RATH, G.J.   196

"Childrens Choice of Playground Equipment: Development of Methdology
for Integrating User Preferences Into Environmental Engineering,"
Journal of Applied Psychology,   Vol. 58,   No. 2, October, 1973, pp.
233-238.

KEY WORDS:

Man	Setting	Relational Concepts	Outcomes
Children	_Playground_	_Equipment pre- ferences_	_Scaling technique/ design_

The study was aimed at determining whether preference scaling techniques could be used as a means of aiding engineering systems design.
Children's play equipment was chosen as the design area. Engineering
design and the development of technology, in general, is often carried
on without systematic and operational involvement of the ultimate users
of these systems.

The study tested three hypotheses: 1) children can construct
internally consistent preference scales for play equipment using photographs as stimulus material; 2) the preference scales will be consistent over time; and 3) the preference scales so constructed will be
correlated with actual use of the equipment in a playground setting.
Paired comparison and rank-order methods were used to determine the preferences of 48, eight and nine year old children.

Photographic stimuli consisted of 5 x 7 color prints that were
mounted for presentation. The six pieces of equipment photographed
were: a) monkey bars; b) slide; c) high-low (horizontal) bars; d)
see-saws; e) horizontal ladder; and f) swings. These third and fourth
grade children were shown the photographs and asked to demonstrate their
preferences by two methods mentioned above at two points in time (4
month intervals). In all cases, the subjects had little difficulty with
either of the methods. The children were quite responsive to the tests
and appeared to be highly motivated. The interviews were approximately
15 minutes in length.

During the months of August and September, 1970, a time-lapse movie
camera was set up to unobtrusively record the play behavior in the
playground where the interview photos were taken. Filming was done on
12 occasions. These measures were taken to allow estimation of the
probability that a given piece of equipment was preferred over others by
its level of usage. Probabilities estimated in this manner were then
compared with data derived from the internal scales. Major problems
encountered through used of this method were that: 1) it was often difficult to determine whether a child was using the apparatus for the
function that it was designed for; and 2) in many frames, there was no
way to determine if a child was using one of the devices or merely
observing use of it. The types of behaviors observed via the films were
interval scaled and analysed.

The results indicated that young children can produce internally
consistent preference scales for play equipment using these methods, and
the preferences were correlated with actual usage of the quipment in a
physical setting. Predictions justified by the photo-interviews were

not contradicted by any of the scales predicted from the behavioral observations. For those preference relationships where the two measurement approaches agreed, the photo interview method was less sensitive than the behavioral observations, and observed differences in equipment usage were quite small. The results also suggest that scaling techniques can be used as part of the design process.

PLANEK, T.W. AND FOWLER, R.C.   197

"Traffic Accident Problems and Exposure Characteristics of the Aging Driver," _Journal of Gerontology,_ Vol. 26, No. 2, April, 1971, pp. 224-230.

KEY WORDS:

Man	Setting	Relational Concepts	Outcomes
Elderly drivers	_Urban-rural; two states_	_Frequency; exposure; context_	_Differences; judgements_

This paper emphasized that educating the aging driver should be based on perceptual and accident-related violation problems that are particularly acute for that person's age group. The many environmental variables that influence driving behaviors need to be taken into consideration when assessing the situational aspects of accidents involving the elderly. The authors stressed the need for various agencies to go beyond molar estimates of annual mileage in specifying safe driving rates since a wide array of variables interplay which can drastically differ from state to state.

A questionnaire was distributed during 1968 which was based on data supplied by the state of Illinois, the California Department of Motor Vehicles, and the American Association of Retired Persons (AARP). This research instrument included questions concerning biographical backgrounds, driving experience, and various other factors. Data was also obtained from two states that differentiated between urban and rural accidents. The hypothesis was that there would be no difference in the probability of having a given type of violation between the over 55 and under 55 age groups. The four violation types accounted for differences in the environmental context of the accidents (data from two states, urban-rural differences).

Forty eight percent of the questionnaires were returned; of these, 48.1%, 43.8%, and 52.6% were from Illinois, California, and the AARP, respectively. Analysis of data indicated that responses of the three samples differed on almost every demographic item and on many of the driver experience items. Definite trends in response patterns across all groups were discernible.

In the measure of driving frequency, it was reported that males generally drove more than females, and drivers under 65 indicated a higher frequency than those of the same sex, 65 and older. Rush hour driving frequencies indicated several trends, particularly between rush-hour and rural driving. The amounts of night time and winter driving done by respondents differed from state to state, due to climate variations and males' social needs and habits. Respondents were asked to estimate the percentage of their total driving mileage on four types of roads: a) expressways; b) highways and rural roads; c) main streets; and d) residential streets. The two most frequently traveled road types for both males and females were types b and c. Three groups of drivers emerged from these analyses concerning frequency of exposure to specific conditions, both environmental and otherwise. Commuting to and from work comprised the bulk of individual driving necessities.

In conclusion, these efforts towards understanding the aging driver presented an informative approach that dealt with the constantly expanding contingency of age 55 and over drivers in the U.S. The typical driver perceived himself as driving too slowly, while also following other vehicles too closely. It appeared that these drivers were not troubled by problems of perceptual judgement nor maladaptive responses when driving in normal traffic conditions in either urban or rural locations. The authors stressed that the need to cope with excessive traffic volumes, unfamiliar routes, and high speed driving should be minimized, particularly after age 65. Future research efforts need to assess the effects of stimulus overload on elderly drivers.

POPENDORF, W. AND SPEAR, R.   198

"Preliminary Survey of Factors Affecting the Exposure of Harvesters to Pesticide Residues," _American Industrial Hygiene Association Journal_, Vol. 42, No. 7, June, 1974, pp. 374-380.

KEY WORDS:

Man	Setting	Relational Concepts	Outcomes
Harvesters	_Fields_	_Pesticides/ exposure_	_Dust exposure level_

This study focused on post-exposure effects of agricultural field workers to organophorous pesticide residues.

The authors developed a graphic description of the various environmental, toxiological and intervening factors influencing the exposure to and absorption of pesticide residues by the agricultural workers when tending to fruit crops. A subjective survey was conducted to obtain some preliminary information of the work procedures and working conditions of harvesters which might affect their exposure to pesticide residues.

This study presented results of the survey and served to illustrate the considerable differences in work practices and the resulting difficulty in quantifying pesticide exposure after effects.

The survey was conducted in the central valley of California during late August, 1972. The crops observed included cling peaches, oranges, and grapes. These crops all are associated with a past history of residue poisoning by organophosphate pesticides among harvesters in California.

The work crews in the survey were selected by chance encounter during an inspection tour of selected areas. Observations were recorded on the thermal environment, clothing, work practices, and the hours of workers. Subjective estimates were made of potential pesticide exposure via skin and respiratory routes.

The ages of the workers were estimated by observation; four percent were verified by direct interrogation. The sex distribution of the crews were also recorded. Family composition was observed, with many families specializing in picking certain crops.

The clothing worn by the pickers was also observed. Long sleeve shirts predominated. This gave protection to the pickers from the rough branches and thorns they encountered while working. Gloves, however, were not worn by those picking grapes or peaches.

Airborne dust samples were taken on a close 37mm Millipore cassette (0.8u) using Cassella personal air samplers at flow rates near 1.5 liters per minute.

The sampled pickers were typical of harvesters and harvesting practices in the central valley of California. In this locale, it appears that worker exposure to pesticide residue is due principally to the dislodging of pesticide-containing dust from the foliage by the worker during the harvesting of these crops.

POULTON, E.C., HUNT, J.C.R., MUMFORD, J.C. AND POULTON, J.   199

"The Mechanical Disturbance Produced by Steady and Gusty Winds of Moderate Strength: Skilled Performance and Semantic Assessments," *Ergonomics,* Vol. 18, No. 6, 1975, pp. 651-673.

KEY WORDS:

Man	Setting	Relational Concepts	Outcomes
Women	*England*	*Wind effects*	*Task decrement*

The purpose of this study was to determine maximum acceptable wind speeds for pedestrians. The authors hypothesized that measurements of acceptable wind speeds could be obtained in semantic as well as psychological tests.

A wind tunnel was used as the experimental setting in which wind velocities of 4 and 8.5 meters per second (m/sec) were tested. Turbulence was produced by vanes within the tunnel. Levels of disturbance were tested at .5% and 12% turbulence. Air temperature and lighting were held constant for the duration of the different environmental settings. Two groups of 11 women were filmed in the tunnel while performing a series of nine different experimental tasks. These tasks were walking, dressing, manual dexterity, and visual acuity in a variety of differing wind speeds and degrees of turbulence.

It was shown that increasing the wind velocity from 4 to 8.5 m/sec measurably decreased the effectiveness of all the task variables and increased the time required to perform the various test situations. It was found that the addition of the 12% turbulence was almost as detrimental to the performance of tasks in the 4 m/sec wind as was the 8.5 m/sec velocity condition. The semantic assessments of the task performances was similar to the actual test situations. The correlations between individual subject performance and their verbal assessments were not different from that expected by chance. The authors concluded that semantic responses were based upon the subjects' expectations rather than what was actually occurring during the test situation.

PUTZ, V. AND K.U. SMITH   200

"Human Factors in Operating Systems Related to Delay and Displacement of Retinal Feedback," *Journal of Applied Psychology,* Vol. 55, No. 1, February, 1971, pp. 9–22.

KEY WORDS:

Man	Setting	Relational Concepts	Outcomes
Machine operators	*Simulated work environment*	*Retinal feedback*	*Movement capability*

A central question in human factors theory concerns man-machine relationships and the way in which operators correct for and adapt to spatial non-compliances in feedback between eye, head and manual movements, and visual input. Therefore, the nature of response to displaced and delayed vision also represents a theoretical issue for human factors theory of machine operations. Theories of response to altered retinal feedback - i.e., associative learning doctrines and the feedback-compensation hypothesis were evaluated relative to their application in defining human factors principles in machine and perceptual training designs.

Using 12 subjects, controlled comparisons were made of the relative effects of reversed and delayed feedback of head and eye movements under conditions in which head movements could not compensate altered feedback of eye movements and visa versa. Accuracy of ocular tracking was affected more by reversal of retinal feedback of eye movements than by head tracking which was changed by reversal of visual effects of head movements.

Eye tracking was reduced to highly irregular, unstable vision, whereas saccadic head movements retained a relatively high level of tracking accuracy. Head-eye tracking combined the effects of reversal of feedback that were found with each of the specific types of movement. Delay of retinal feedback of the movement systems at intervals of 0.0, 0.2, 0.4, 0.6, 0.8, and 1.5 sec. caused progressive elimination of unstable ocular guidance, but also increased the error of such tracks.

Other results indicated that there was little or no learned adaption to the reversed and delayed vision produced by head and eye movements. The findings give support to a behavioral cybernetic interpretation of the guidance factors in non-machine and perceptual systems relationships by showing that the effects of altered feedback in

machine and systems operations are determined by movement capabilities
in compensating displacements and delays in sensory input. The results
also suggest that visual impairments may be produced by delays in the
retinal feedback effects of eye and head movements and that these
defects may require dynamic methods of optometric diagnosis and training
for their measurement and correction.

RAE, A. AND R.M. SMITH   201

"Subjective Odor Levels in an Air-Conditioned Hospital Ward," *Applied
Ergonomics,* March, 1976, 7(1): 27-34.

KEY WORDS:

Man	Setting	Relational Concepts	Outcomes
Hospital patients	*Surgical ward*	*Ventilation rate*	*Odor perception*

     In order to maintain a high level of sanitation and sterility,
hospitals are forced to use closed circuit air-conditioning systems
which limit the amount of fresh air that enters the patient areas. The
authors propose that this recirculation of air does not affect the subjective odor levels of the patients and will not vary with a sizeable
increase in the recirculation rate.

     A general surgical ward which accomodates 23 male and 23 female
patients in East Kilbride, England, was used as the experimental unit.
The ward supplied its air-conditioning through a central plant and could
maintain a recirculation rate of air betwen 0 and 80 percent.
Mechanical air changes were varied during the test from six down to
three per hour with either 0 or 80 percent recirculation rates being
tested for each air change range. Four interviewers questioned approximately 50 percent of all ward patients concerning perceived odor levels
as well as a series of other indices of satisfaction such as privacy and
noise levels as a check for overall complaint levels. Odors within the
ward were normally fairly bland, although experiments were conducted
during periods of acute odors caused by putrifaction in patients.

     The investigators found no significant variances in perceived odor
levels between the high air change of 6 per hour with 0 percent recirculation (6.7 percent odor complaint rate) and the low of 3 per hour
with 80 percent recirculation (9.0 percent rate). Even during the two
occurences of the acute levels, only a small increase in the complaint
rate was found, this being primarily related to proximity to the source
of the odor rather than the rate of ventilation. It was the opinion of
the researchers that the lowest rate of air change (3 per hour with 80
percent recirculation) was a perfectly adequate level based upon perceived odor levels.

"Environmentally Triggered Thrombophlebitis," *Annals of Allergy,* August
1976, Vol. 37, No. 2, pp. 101-109.

KEY WORDS:

Man	Setting	Relational Concepts	Outcomes
Thrombo-phlebitis patients	*Hospital*	*Treatment settings*	*Health management*

The relationship of environmental factors to the onset and control
of human illnesses has long been a major concern of the clinical
investigator. Phlebitis is one disease with which little is currently
known regarding its triggering agents. This clinically controlled study
was designed to gain insight into potential environmental methods of
thrombophlebitis control.

Ten patients with recurrent non-traumatic thrombophlebitis were
randomly selected to participate in the study. Detailed case histories
of the one male and nine female subjects, ages 26-66, were then
gathered. The study consisted of a diagnosis and treatment period
totaling at least sixteen days.

Specially constructed rooms served as the treatment environment.
They were designed and maintained to control against the existence of as
many petrochemicals as possible. Filters were utilized to eliminate
extraneous fumes and odors that might come through the doors. No medi-
cations were given during the patients' stay in their rooms, except for
bicarbonate of soda. As many as thirty types of source foods, which
were chosen for their relative absence of petrochemicals, were used in
the treatment part of the study. The patients were later retested
using the same types of food, but this time obtained from the commercial
market and containing artificial preservatives and additives. Other
tests were implemented which exposed the patients to such things as
natural gas, cigarette smoke, perfume, pine scented floor wash, and
ethyl alcohol.

The results showed that "all patients, once under environmental
control, cleared their active symptoms, including their onging
phlebitis, in three to ten days." It was found that many individual
differences existed between patients, and that some individual stimuli
would reproduce all of the patient's original symptoms, while others
reproduced some of them. Yet, regardless of the differences, upon
release from the treatment setting the patients were able to manipulate
their environments at home so as to eliminate the environmentally
induced triggering agents, and consequently remain phlebitis-free.
Based on this knowledge, the author concluded that "environmentally
triggered phlebitis can be cleared and reproduced by the institution of
rigid environmental control."

REA, W.J.   203

"Environmentally Triggered Cardiac Disease," *Annals of Allergy,* April,
1978, Vol. 40, No. 4, pp. 243-251.

KEY WORDS:

Man	Setting	Relational Concepts	Outcomes
Cardiac patients	*Hospital*	*Treatment setting*	*Health management*

Evidence of environmentally triggered non-arteriosclerotic heart
disease has rarely been reported. The author's experience with cases
which apparently involved environmentally induced cardiac arrhythmias
suggested the need for further examination of the subject.

Twelve patients aged 12-45 possessing unexplained cardiac arrhythmias were studied. After detailed ecological histories were taken, the
patients were placed in specially constructed rooms which were designed
to create an environment that was as contaminant-free as possible.
Patients were kept under this tight environmental control for at least
sixteen days. Food was chosen for its relative absence of petrochemicals. At a later time, testing was done with foods which had preservatives and additives in them. Odors and fumes which are commonly found
in the home and work environment were pre-tested and administered to
patients.

Upon completion of the controlled-environment study, ten of the
twelve patients cleared their arrhythmias and related symptoms in three
to ten days. When foreign substances were ingested,"ninety-five percent
of the reactions occurring within the first four hours started within
the first five minutes after ingestion, leaving no doubt in the minds of
the observing personnel and patients that there was a direct cause-and-
effect relationship." Reactions to inhaled chemicals showed similar
results. Thus, the author concluded that cardiac disease can be induced
and controlled by the environment. The significance of these results is
that once the triggering agents have been identified, precautions can be
taken in the home and work environment to diminish the incidence and
severity of certain types of cardiac diseases and their accompanying
symptoms.

REES, C.A.   204

"Age Segregation in Nine Cities," *Proceedings of the Association of
American Geographers,* Vol. 7, 1975, pp. 55-59.

KEY WORDS:

Man	Setting	Relational Concepts	Outcomes
Elderly	*Urban*	*Size and growth rate*	*Age segregation*

As American culture has become more youth-oriented over the years, being over the age of sixty has assumed negative connotations. The age grading process has influenced social stratification in many ways. This study investigated age segregation and the spatial parallel to age grading.

Taeubers' index of residential segregation, centrographic measures, and entropy statistics from nine cities for the years 1960 and 1970 were used to measure the concentration and segregation of the elderly. The cities were selected for variety of size and region. Data were collected on census tracts which had identical boundaries for the two periods. Taeubers' index showed that residential segregation for persons aged 65 and over increased between 1960 and 1970. For persons in the middle age group, (55-64), segregation decreased. The degree of segregation was not as extreme as the segregation of racial minorities. The two cities whose elderly population became less segregated, Altoona, Pennsylvania, and Witchita, Kansas, were also the only two cities studied which lost population. Though centrographic measures revealed a pattern of age segregation, the interpretation of results was more tenuous. The elderly were slightly more concentrated, though dispersion rates of the two older groups were slightly greater than the dispersion rates for groups under the age of 55. Entropy statistics showed that even though the elderly were the most concentrated they became more dispersed in six of the cities over the ten year period. Patterns of concentration and segregation were both positively related to growth rate. The Taeuber indices were postively related to both city size and growth rate as was dispersion of the elderly as demonstrated by the entropy statistic.

Since most cities grow in "concentric" rings of housing and settlement, new housing tends to be occupied by persons in roughly the same age group, with each successive age group being more spread out than its predecessors. Although the elderly are becoming more dispersed they are thus simultaneously becoming more segregated by the bracket effects of each succeeding ring of settlement. Though age segregation may result from the transient effects of rapid growth areas, it is becoming institutionalized by current practices in building, planning, and land use regulation. The physical segregation presently experienced may have adversely affected the interaction necessary to maintain social integration of the elderly with the balance of the populations. At the scale of the city age segregation could produce similar negative effects on local neighorhoods as does social segregation with its accompaning low status and low incomes.

REICHNER, R.F.   205

"Differential Responses to Being Ignored: The Effects of Architectural Design and Social Density on Interpersonal Behavior," _Journal of Applied Social Psychology,_ Vol. 9, No. 1, January-February, 1979, pp. 13-26.

KEY WORDS:

Man	Setting	Relational Concepts	Outcomes
College freshmen	Dormitories	Spatial arrangement	Social isolation

It is proposed that certain architectural arrangements of space have the effect of altering an individual's desired level of privacy. The author hypothesized that groups of people living in either semi-private or communal living arrangements will behave differently when exposed to a series of social encounters.

Two sets of experiments involving 32 freshmen from the State University of New York at Stony Brook were performed to measure the effects of being ignored in conversation in relation to their environmental surrounds. Sixteen of the students occupied a room in a standard, double-loaded college dormitory while the other 16 were residents in suite arrangements in which four occupants shared a common bath and study facility. It was hypothesized that when systematically ignored in a test conversation between two or more individuals, the residents of the suite arrangement would feel much more uncomfortable and rejected than those who were used to sharing large common facilities in the corridor dorm.

Results of both experiments confirmed the study hypothesis that those students living in the suite design were approximately 30 percent more likely to feel a sense of isolation in stressful social encounters than those students in the more traditional corridor dorms. It was proposed that corridor residents developed responses which limited their social involvement in an attempt to reduce the number of unwanted encounters. Suite residents, on the other hand, were used to the constant interplay among small group behavior and felt threatened by exclusion from interpersonal contact.

RICHARDS, O.   206

"Effects of Luminance and Contrast on Visual Acuity, Ages 16 to 90 Years," *American Journal of Optometry and Physiological Optics,* Vol. 54, No. 3, March, 1977, pp. 178-184.

KEY WORDS:

Man	Setting	Relational Concepts	Outcomes
16 to 90 year olds	*Room*	*Luminance & visual acuity*	*Optimal levels*

This study was performed to acquire information on how well people can see at night-driving luminances. Because reading warning signs quickly is often the critical task in automobile driving, letter tests were determined to be especially appropriate. Subjects aged 16 to 90 years were tested at 10 fl. to ensure that a subject's vision was within the normal range at 0.1 fl., representing average road luminance at night, and at 10 times and 1/10 of this average luminance. Target contrast and size were varied to reveal their importance in the decrease of legibility with decreasing light and increasing age.

Almost all subjects except the very oldest were automobile drivers. The very old subjects were of better than average ability for their age. No obvious visual pathology was manifest in any of the subjects, although no attempt was made to evaluate eye transmissions.

This study suggests that gains in older people's vision could be expected to result from increased lighting. Vision acuity does decrease with age and this quite naturally affects driving. Fortunately most elderly people give up driving at night when seeing becomes difficult for them. No single age can be suggested for prohibiting night driving because of the great variation in vision loss from aging. A possible solution might be to prohibit such driving when the licensee can no longer see an object of given substance and contrast.

The author suggests that well-lighted highways of 1 fl. or more would reveal objects of 5 min subtense and 11% or more contrast for most drivers. Brighter lighting than that would not greatly increase acuity except possibly at higher speeds and for the oldest drivers.

RILEY, R.L.   207

"The Ecology of Indoor Atmospheres: Airborne Infection in Hospitals," *Journal of Chronic Diseases,* Vol 25, No. 8, August, 1972, pp. 421-423.

KEY WORDS:

Man	Setting	Relational Concepts	Outcomes
Hospital patients	*Hospital*	*Airborne infection; indoor air pollution*	*Reduced infection, air disinfection device*

The author discussed research pertaining to the problem of indoor air pollution in hospital environments, which in his opinion is a threat to human health more immediate than outdoor air pollution. It was maintained that indoor environments do not provide the vast dilution of noxious materials which occurs outdoors.

Air movement within the enclosed atmosphere of a hospital conveys an array of potentially infectious air particles to all portions of the building. The organisms that flow through heating, ventilation, and air conditioning systems (HVAC) together with the air transport capacities and functions of corridors, stairways, and elevator shafts are potentially infectious until they die, are vented to the outdoors, or are killed. The consequences of these occurences are particularly unfortunate in hospitals because of the "hypersusceptible and hyperinfectious clientele." Respiratory infections, in particular, are extremely affected by the contents of the air within the hospital.

The author states that the most appropriate control measure is to abate the flow of airborne infection via air disinfection methods. This has been studied at the John Hopkins University School of Hygiene and Public Health. These studies have found that upper air irradiation with ultraviolet (UV) lighting can rapidly disinfect the air of a room, and corridor irradiation can minimize the spread of potentially harmful airborne organisms through a building. It was found that the original amount of infectious airborne organisms were reduced to 37% of the initial concentration using UV fixtures that were suspended from a ceiling in the middle of a hospital room. The device was left on around the clock. The upper and lower air convection currents interacted without

was required and the outbreak of diseases such as tuberculosis have sub-
sided at the institution cited above since the units were installed and
observed.

It was acknowledged that very little literature exists which
focuses on the relation between indoor air pollution and infectious
diseases in hospitals. Findings from studies focusing on this problem
were briefly discussed. The author concluded that solutions to this
environmental health problem are more likely to succeed in hospitals as
opposed to schools due to one major advantage: patients stay within the
hospital building 24 hours a day while school children are in school for
only 5 or 6 hours per day. There was little evidence that the UV units
are capable of abating the flow of infectious organisms transmitted in
non air-related manners. This method was not viewed as a complete
answer to the problem of hospital acquired infection, although the
method was discussed as a possible contributor to preventive medicine
and to a better understanding of the "ecology of indoor atmospheres."

RODIN, J. AND E.J. LANGER   208

"Long Term Effects of a Control-Related Intervention with the Institutionalized Aged" *Journal of Personality and Social Psychology,* 1977,
Vol. 35, No. 12, pp. 897-902.

KEY WORDS:

Man	Setting	Relational Concepts	Outcomes
Elderly	*Nursing homes*	*Control; respon-sibility*	*Lower mortality rates*

The effects of an intervention designed to encourage elderly
nursing home residents to make a greater number of choices and to feel
more in control of day-to-day events was assessed by this study. It was
intended to determine whether the decline in health, alertness and activity that generally occurs in the aged in nursing homes could be slowed
or reversed by choice and control manipulations.

Ninety-one subjects took part in the study. Half of the groups
were given a talk by a hospital administrator emphasizing their responsibility for themselves; communication to a second, comparison group
stressed the staff's responsibility for them as patients. To bolster
the communication, residents in the experimental group were offered
plants to care for, whereas residents in the comparison group were given
plants that were cared for by the staff.

Data collected included nurses' ratings of each patient on a semantic differential scale for mood, awareness, sociability, mental attitude, and physical activity; physicians' ratings of medical records for
periods before and after the intervention; and attendance at a lecture
on psychology of aging. The number of people from each group attending
the lecture and the frequency with which they asked questions was
recorded. Mortality rates during an 18 month period was obtained and
compared to previous rates.

The most striking data were obtained in death rate differences between the two treatment groups. Taking the 18 months prior to the original intervention, it was found that the average death rate during that
period was 25% for the entire nursing home. In the subsequent 18 month
period following the intervention only 7 of the 47 subjects (15%) in the
responsibility-induced group died, whereas 13 of the 44 subjects (30%)
in the comparison group died. Residents in the responsibility-induced
condition also showed higher health and activity patterns, mood and
sociability.

ROHNER, R.P.   209

"Proxemics and Stress: An Empirical Study of the Relationship Between
Living Space and Roommate Turnover," *Human Relations*, Vol. 27, No.7,
September, 1974, pp. 697-702.

KEY WORDS:

Man	Setting	Relational Concepts	Outcomes
Male students	*Dormitory*	*Crowding*	*Stress*

Prior studies have suggested that crowding in confined spaces will
cause social tensions among occupants and lead to the breakdown of
interpersonal relations. The present study tested the hypothesis that
crowding in college dormitory rooms will lead to the turnover of roommate pairings.

Fifty-eight pairs of roommates in four men's dormitories were used
as the sample population of this study. Thirty-seven of the rooms had
twin bed furnishings while 21 had bunk bed sleeping arrangements. Since
all rooms were of the same total area (159 square feet) the bunk bed
arrangement allowed approximately 20 percent more available living space
because only one bed actually occupied floor space. Satisfaction and
interpersonal tensions were measured by comparing roommate composition
during the spring term of an academic year and again after the 116 members had returned in the fall. All room assignments in the fall were
voluntary and a matter of discretion between each room pairing.

The author found that a turnover rate of 54 percent had occurred in
the twin bed rooms while only 24 percent of those occupying bunk bed
rooms declined to share a room the following fall term. Although the
total area of all rooms was less than the spatial optimum of 215 square
feet for two people, the crowding created by the addition of the extra
bed space in the twin bed room seemed to be a significant factor in
causing the higher turnover rate. The author also suggested that those
living in the bunk bed arrangements were accorded a greater degree of
visual privacy. This finding also explained the lower frequency of
roommate tension, although acoustical separation was equal for both
types of beds.

RONALD, P.Y., M.E. SINGER AND F. FIREBAUGH   210

"Rating Scales for Household Tasks," *Journal of Home Economics,* Vol. 63,
No. 3, March, 1971, pp. 177-179.

KEY WORDS:

Man	Setting	Relational Concepts	Outcomes
Homemakers	*Residences*	*Task dimensions*	*Preferences*

Rating scales were discussed relative to their application in the
residential environment and how a user's daily and long term needs and
desires might be better understood than they have been in the past. The
use of rating scales in the behavioral sciences was viewed by the author
as a development that can be of great use to the home economist. A
scale to measure attitudes of homemakers towards their work was reported
in 1962 (F. Maloch) and the follow-up study reported here sought to provide a comparative basis for further research.

Rating scales in this study: 1) ascertained characteristics of
household tasks which were most and least liked by persons; and 2)
determined the relevance of each scale in another setting and with different subjects. The same rating scale was administered to two similar
subject groups, one in Canada (N=63) and one in the United States
(N=120). Personal interviews were conducted with each subject and a
questionnaire was completed, including a question asking the person to
name the task that was most liked, and in turn, least liked. These two
tasks were each given a 5-point scale and subjects placed a check in the
most appropriate blank for each task. Mean scores were calculated. One
point was used for the positive end of the scale and 5 points for the
negative ends.

Results indicated that in both studies cooking was liked best by
subjects. Ironing was least liked in the present study, while cleaning
was least liked in the earlier study. Each subject provided a set of
reasons for stating their preferences. More than 30 percent in each
study reported pride in results, satisfying work, and appreciated
efforts by family as principal rationale for naming their most liked
tasks. In turn, short-term results, dislike for time spent (in tasks),
monotonous, and tiring attributes were most associated with least liked
tasks. The similarity in findings for the most and least liked tasks
cut across both groups. Also, specific characteristics of tasks and a
person's selective rationale cut across task ratings.

Characteristics of tasks most frequently mentioned are items that
were seen as deserving the most attention from researchers. Although
the dimension of monotony has been studied in actual industrial
environments, this dimension has not been addressed in home environment
research.

It was concluded that: 1) certain scales need to be retained for
further research; 2) larger random sample groups are needed; and 3)
multivariate analysis should be performed on data in future studies concerning rating scales for tasks in the home environment.

"Intransitivity of Preferences: A Neglected Concept," *Proceedings of the Association of American Geographers,* Vol. 7, 1975, pp. 185-189.

KEY WORDS:

Man	Setting	Relational Concepts	Outcomes
College students	*Laboratory*	*Intransitivity*	*Environmental preferences*

     Much attention in recent literature on the nature of spatial pre-
ference has been directed to individual differences in cognitive organization. A variety of preference functions may be reflected in individual responses. In an attempt to identify structural consistencies, many
decision theories have included the principle of transivity. This principle can be defined by examining a series of dichotomous choices; if A
is preferred to B and B is preferred to C, transitivity infers that A is
preferred to C. This assumption of transitivity is thought to contribute to the inconclusive results obtained in recent preference studies.
This study was developed in an effort to study intransitivity as an
independent component of residential preference. Two choice theories
and their relationship to the transitivity maxim were examined.

     Ninety-four undergraduate students were the subjects in a paired
comparison study. Twenty residential attributes, divided equally between site and situational factors were presented in a randomly ordered
paired comparison format. Site factors included size of lot, privacy,
interior and exterior appearances, rental or purchase value, and single
family housing. Examples of situational factors were social prestige,
proximity to employment or commercial areas, quality of schools and
racial integration. Principal component analyses were performed on the
frequencies of selection. One week later eight hypothetical neighborhoods representing five attributes each were presented to the same
group for ranking. Results from the paired comparison task were used to
prepare "expected" rankings to compare the actual rankings by the
subjects. The "expected" rankings were formed under the assumption of
transitivity. The two sets of rankings were compared using Kendalls
rank correlation. Results were inconclusive. Close correspondence and
little or no correspondence were found with equal frequency.

     The two choice models (Tersky, 1969) were introduced in an effort
to include measures of intransitivity. Under each of these models the
gross amount of intransitivity was calculated and treated as an independent variable in a simple bivariate regression. Use of the additive
model in such a manner considerably improved the veridicality between
"expected" and "observed" rankings. Use of the second more complex
additive differences model could not be recommended. A large loss of
parsimony offset the slight increase in explanation. Several methodological ramifications of the transitivity problem are discussed.

ROTHBLATT, D.N.   212

"Housing and Human Needs," *Town Planning Review,* Vol. 42, No. 2, April,
1971, pp. 130-144.

KEY WORDS:

Man	Setting	Relational Concepts	Outcomes
Adults/general population	*Public housing*	*Human needs/ housing*	*Adjustment*

In 1963, a modest sample of the adult population living in Marlboro
Houses, a public housing project in Brooklyn, New York, participated in
a survey in order to acquire knowledge concerning family needs and
housing.

Two separate groups were involved: 1) families living in seven
story buildings, and 2) families living in 16 story buildings with a
communal open terrace on each floor.

Personal interviews, in the form of questionnaires, were used with
40 families at each site. The interviews, conducted through chance
meetings on the site, lasted from 1 1 1/2 hours. Eighty-nine percent of
those interviewed were white and all were female heads of household.

The survey included the following human need variables: 1. family
needs; and 2. belongingness needs; 3. esteem needs; and 4. independence
needs. The first variable included issues focusing on safety for children, shared activities, and the husband's participation in activites.
Variable two concerned friends and the person's sense of accomplishment.
Esteem need concerned persons pride in home and building, status with
friends and husband's sense of accomplishment. Independence needs
involved inward personal privacy, size and uniqueness of apartment.

The results of the survey indicated that the 16 story buildings
with terraces on each floor were more likely to fulfill the family and
belongingness needs although seven story buildings made for more
independence. Families living on the lower floors of either type of
building gained "lower floor" superiority with the family and belongingness needs responded to in the most positive manner.

These results would suggest that the information system (the organization) should relate needs, life style and preferences of different
client groups to environmental factors such as building types, interior
space, private and communal outdoor space, and social and physical
features of the community at large. The author concluded that only then
will most of the human needs in such settings be satisfied in a proper
manner.

"Two Studies of Crowding in Urban Public Spaces," *Environment and Behavior,* Vol. 7, No. 2, June, 1975, pp. 159-184.

KEY WORDS:

Man	Setting	Relational Concepts	Outcomes
Shoppers/ commuters	*Department store, terminal*	*Crowding, stress, cognition*	*Low performance, crowded conditions*

Two studies were designed to investigate the hypothesis that: 1) increased numbers of people occupying a constant area increase the complexity and uncertainty of the environment, leading to an inability on the part of occupants to gain a clear environmental image; and 2) such an overload can lead to task decrements and negative effect for persons in the settings who are required to perform tasks requiring understanding and manipulation of the environment. Based on prior studies, it was expected that regardless of the task involved, the intensity of affect would be increased by higher density.

The setting for Study I was a section of a large Manhattan department store. The emphasis was on the problem of attaining cognitive clarity in a crowded situation. Individual subjects were exposed to either a crowded or uncrowded condition and a series of recall tasks were performed. It was found that subjects tested under crowded conditions performed similarly to uncrowded counterparts on the focal tasks. However, cognitive maps drawn of the area by crowded subjects were less accurate and complete. Crowded individuals provided more positive descriptions of sale items in the store than uncrowded subjects.

In Study II, subjects were given a series of tasks to perform in a busy Manhattan railroad terminal. Tasks were similar to those normally encountered by persons in this setting. Knowledge of manipulation, and movement through the setting was required. Half of the subjects were provided with a map of the setting, while half were not. All subjects had to perform the same set of tasks. Results indicated that: 1) crowded subjects performed fewer tasks; 2) felt more negative about their performance and the physical and social environment; and 3) performed differently on a post experimental task previously used as an indicator of stress. Also, sex by density interactions occurred. The implications of the results for both planners and architects were discussed.

"A Note on the Relationship Between the Comfortable Interpersonal
Distance Scale and the Sociometric Status of Emotionally Disturbed
Children," *Journal of Genetic Psychology,* Vol. 128, March, 1976, pp.
91-93.

KEY WORDS:

Man	Setting	Relational Concepts	Outcomes
Children	*Institution*	*Interpersonal distance, socio-metric status*	*Preferences*

Relationships were investigated between two measures of social
attraction or status via the Comfortable Interpersonal Distance (CID)
Scale and a traditional measure of sociometric status. Both measures
were recorded for a sample of 47 preadolescent boys (ages 8-13) who were
emotionally disturbed. All subjects were in residential treatment.

The CID device involved marking on a diagrammatic drawing of a room
how close a subject wanted other people to sit from him or her in the
room in order for the subject to be psychologically comfortable. The
purpose of the study was to supply preliminary evidence concerning the
validity of the CID with a preadolescent population. This measure was
studied together with a sociometric questionnaire to obtain validity
estimates. A series of questions were asked of each boy to ascertain
positive and negative sociometric preferences. It was hypothesized that
the more a child was liked by his peers, a lower CID score would be
evidenced. In these cases the child's peers would want to sit closer to
him or her in the room. Pearson correlation coefficients were computed
between the two testing measures.

The hypothesized relationships between these measures were highly
correlated ($r = -.800$). Thus, the more a boy was liked by his peers, the
closer these peers indicated they would like to be proximate to the
boy in the room.

Physical interpersonal distance was seen as a nonverbal method of
communicating a favorable attitude toward a person. It was concluded
that both the CID and the sociometric technique were measuring the same
variable - social popularity. The advantage of the CID was that subjects are required to think about and rate every other person in the
group, while the sociometric procedure only requires persons to identify
extreme preferences. The CID was seen by the authors as being a valid
and stable measure of social popularity for these children.

SCHIFFENBAUER, A.I., J.E. BROWN, P.L. PERRY, L.K. SCHULACK,   215
AND A.M. ZANZOLA.

"The Relationship Between Density and Crowding: Some Architectural
Modifiers," _Environment and Behavior,_ Vol. 9., No. 1, March, 1977, pp.
3-14.

KEY WORDS:

Man	Setting	Relational Concepts	Outcomes
Female students	_Dormitory_	_Architectural arrangement_	_Perceived crowdedness_

It is suggested by the authors of this study that actual room density of a dormitory as a cause of crowdedness among its occupants is
only one environmental variable and is tempered by a range of other
architectural features. Specifically, they propose that rooms with more
useable space will be judged less crowded, rooms on higher floors will
seem to the occupants larger and more open, and rooms which receive more
sunlight will be perceived as larger and less crowded than darker rooms.

A female student dormitory at Virginia Polytechnic containing 48
double rooms was used as the experimental setting. A study of the floor
plan revealed that, although each room was exactly the same size, the
amount of useable floor space varied because of door swing placement.
The amount of sunlight was less on one side of the building in relation
to the amount received on the other. Also, since the building was a
twelve story tower, the authors could measure the effects of height
above ground level on perceived crowdedness among occupants. A questionnaire measuring perceived crowding and feelings about space were
administered to the 96 occupants, with 75 actually responding and quali-
fying for use in the test.

The results of the questionnaire confirmed the basic premise of all
hypotheses outlined by the authors. Although all the rooms were of
equal density, different feelings of perceived crowdedness were measured. It was found that floor level was a significant factor in perceived size, with higher floors appearing larger to occupants. There
was no significant relationship between floor and crowdedness. Light
rooms were judged to be less crowded than darker ones, but were not perceived to be any larger. Poor placement was found to have an impact on
perceptions of size, but again, did not support the notion that larger
room perceptions would lead to less feelings of crowdedness.

SCHMIDT, D.E., R.D. GOLDMAN, AND N.R. FEIMER   216

"Physical and Psychological Factors Associated with Perceptions of
Crowding: An Analysis of Subcultural Differences," _Journal of Applied
Psychology_, Vol. 61, No. 3, June, 1976, pp. 276-289.

KEY WORDS:

Man	Setting	Relational Concepts	Outcomes
Three ethnic groups	_Urban area_	_Crowding/density_	_Perceptions/attitudes_

Previous studies are cited that view a number of personal, social, and situational factors from the perspective that the psychological impact of density on human activity may determine the individual's evaluation of the external environment. In this study, a field experiment was conducted to determine if a number of psychological and physical factors that had been previously associated with perceptions of environmental crowding differed with the cultural characteristics of urban residents. Six hundred ninety-seven subjects living in Riverside - San Bernadino, California were surveyed to ascertain perceived crowding in the residence, neighborhood, and the city.

A second goal was to determine if certain density characteristics of the city affect these groups to the same extent or in the same way. It was hypothesized that the content and importance of factors associated with these perceptions will vary greatly across the three subsamples tested (members of Black, White, and Chicano groups).

A questionnaire was developed and administered (83 items) to measure variables that were believed to mediate perceptions of crowding. The major issues studied included: 1) psychological items; 2) perception-of-crowding; 3) demographics; and 4) physical statistics and measurements (proximate data). The subjects were individually interviewed by trained personnel.

A stepwise multiple-regression analysis yielded intercorrelations between levels of perception of crowding for each racial group. Perceptions of crowding at each urban level were the criterion variables, and the physical and psychological measures were the predictors. This was done separately for each of the three groups.

It was found that psychological factors indicative of the impact of physical conditions on the individual provided the best explanation for the perceptions of white subjects. Black and Chicano groups, however, tended to view crowding at each of the analysis levels in terms of the total urban "gestalt" associating physical measures beyond their implied impact. City-related conditions tended to permeate all levels of analysis for the two non-white groups, while they were divided between the neighborhood and city analysis levels for the white subsample. Of all the variables studied, culture differences were the most consistent sample division and provided the most accurate explanation of results.

The authors concluded that further work is needed regarding the treatment of environmental crowding as a subset of environmental perceptions, and a careful look at systematic individual/group differences may provide insights for architects and urban designers.

SCOTT, J.   217

"Testing a Housing Design Reference: A Pilot Study," _Architectural Association Quarterly_, Vol. 2, No. 1, January, 1970, pp. 23-31.

KEY WORDS:

Man	Setting	Relational Concepts	Outcomes
Residents	_Urban_	_Design, social theory, life satisfaction_	_Housing assessment criteria_

Emphasizing the frequent gap between architectural design and social theory regarding what constitutes a good physical environment, the author undertook a pilot study on housing in London to determine what criteria should be utilized to assess the success of a housing area. It was based on comparing the following features of the design environment against a range of social criteria:

Specific problems	vs.	Specific satisfaction with the dwelling and environment
Dislikes	vs.	Likes about the dwelling and environment
Limitations upon	vs.	Scope for home life
Insignificance	vs.	Recognition as an individual
Rootlessness	vs.	Social contact between households
Confusion about	vs.	Comprehension of surroundings
Negative outlook	vs.	Positive outlook on life and abilities to communicate, learn, achieve chosen goals and persist in a line of action

Three housing units were chosen based on the anticipation that they would provide a wide range of environmental contrasts. They included 1) a 13 story slab block of maisonettes with internal access corridors; 2) a 6 story block of maisonettes with external access balconies; and 3) terraced houses around paved squares. One architect/planner and three architects (one trained in the United States) rated each unit according to a housing design rating system. The mean scores were averaged together to form a composite score.

Information was gathered from twelve families in the terraced houses and eight from each of the maisonettes. The questions included sections on: 1) privacy, 2) noise, 3) the neighborhood, 4) proximity to services, 5) physical and mental health, 6) interaction with neighbors, and 7) attitudes to the housing environment.

Results indicated that people can have a strong sense of belonging to their specific living environment, while at the same time having little or no sense of attachment to the neighborhoods in which they live. Furthermore, all of the main categories of the housing design were correlated with the interview results in some way. Since this study was not concerned with the superiority of one environment over the other, but rather in forming a basis for rating the success of a housing area, the results were significant. The author believed that this reference did "enable links between the individual dwelling and occupants' responses on housing to be seen which were not apparent from aggregate scores, help to confirm an expected relationship between groupings of dwellings and social contact, and affirm residents' sense of belonging to specific aspects of their environment."

"Effects of Modulated Noise on Speech Intelligibility with Sensorineural
Hearing Loss," _Annals of Otology,_ Vol. 81, No. 2, April, 1972, pp.
241-248.

KEY WORDS:

Man	Setting	Relational Concepts	Outcomes
Middle-aged males, hearing impaired males	Audially controlled room	Noise, hearing loss	Performance

Past research has indicated that the effect of noise on an
individual's ability to understand speech is mainly one of masking.
Consequently, with normal hearing persons intelligibility tends to
improve as the intensity of speech increases in proportion to the
background noise. Yet, some experimental research has also indicated
that this situation is not necessarily the case for individuals with
certain types of hearing loss. The major objective of this study was to
examine the problems that noise might cause for those persons with sensorineural hearing loss.

Twelve men who were identified as having sensorineural loss were
compared to twelve middle-aged men with normal hearing during four one
hour experimental sessions. Speech reception thresholds for normal subjects varied between 0 and 15 dB (ANSI - 1969) while those for the
abnormal group were in the 30 to 40 dB range. Each subject listened to
50 monosyllabic words in varying noise backgrounds. The test words were
recorded along with white noise which was produced by an electronic
noise generator. Four speech-to-noise ratios were employed, nominally
-8, -12, -16, -20 dB. During each session the subject listened to five
word lists, one for each noise modulation rate and one in a continuous
noise background.

Results revealed that although all subjects had poor discrimination
under continuous noise conditions, normal subjects were able to identify
the test words better than those with sensorineural loss under all
experimental conditions. Furthermore, regardless of the noise-to-speech
ratio all of the subjects improved their performances when the modulation rate was decreased. Yet, normal subjects improved more than the
hearing impaired groups, with the differences most pronounced at the -20
and -16 dB speech-to-noise ratios. In short, those subjects with
hearing loss performed worse in noise than would be expected based on
their performance in quiet surroundings. Since hearing tests are
generally given under quiet test conditions, the authors believe that
their results indicated need for "routine clinical measurment of discrimination in noise for those individuals with sensorineural hearing
loss." Otherwise, although it would be possible to predict that
patients with sensorineural loss will perform worse than normal individuals in a noisy situation, there is no way under the present routine
audiometric testing procedure that a clinician would be able to predict
the extent of the hearing handicap.

"Variations in Psychomotor Efficiency During Prolonged Stay at High
Altitude," *Ergonomics,* Vol. 18, No.5, 1975, pp. 511-516.

KEY WORDS:

Man	Setting	Relational Concepts	Outcomes
Males	*Mountains*	*High altitude, Prolonged stay*	*Impaired psycho-motor efficiency*

The objective of this study was to determine the affects on psychomotor skills people experience when exposed to long periods of high
altitude environments. The authors hypothesized that long-term psychomotor deficiencies would be experienced by subjects acclimated to sea
level environments as determined in proceeding tests of short-term
duration.

Twenty-five Indian soldiers from the southern plains of India, were
used as experimental subjects in this study. They were subsequently
stationed in the Western Himalayas at an altitude of 4000 meters for a
period of 24 months. They were tested at a low altitude (200 m), for
speed, accuracy and overall dexterity and again at periods of 1, 10, 13,
18 and 24 months while at the higher altitude. A standardized psychomotor test was devised for this study which mechanically measured the
subjects' abilities to coordinate eye-hand movement.

In support of past short-term tests and the authors' hypothesis, a
significant decline in psychomotor abilities was recorded after exposure
to the 4000 meter environment. Beginning with a mean efficiency score
of 170, the subjects' skills declined in 10 months to a score of 114.
After the 13th month, however, scores began improving until reaching a
level of 145 after 24 months of exposure. Similar readings were
recorded in both accuracy and speed tests, with improvements becoming
noticeable at the 13th month.

It was postulated by the authors that this improvement to original
levels of competence may be due to psychological factors affecting an
acclimatization to higher altitudes as well as to physiological conditioning.

"Leisure Activities in Retirement Housing,"   *Journal of Gerontology,*
Vol. 29, No. 3, May, 1974, pp. 325-335.

KEY WORDS:

Man	Setting	Relational Concepts	Outcomes
Elderly	*Retirement housing*	*Leisure activities*	*Adjustment*

The issue of providing the elderly with adequate leisure facilities and activities in planned residential environments has been of considerable concern of theoreticians, developers, and residents alike. The advantages and disadvantages of various approaches toward accommodating these users have been more speculative than problematic. The principal argument in favor of incorporating recreational facilities and activities in this housing type is that this dimension of residential life may substitute for prior life-cycle activities of residents, i.e. working, traveling, specific hobbies, etc. New roles and relationships emerge through the years that need to be supported through the physical environment.

Two common assertions about leisure activities in retirement housing were studied: 1) that residence in special retirement housing for the elderly will result in more leisure activities than will residence in conventional dispersed housing; and 2) that greater particiaption in leisure activities is associated with a more favorable outlook on life. Two waves of interviews were conducted with 100 residents at each of six retirement housing sites (retirement hotel, three retirement villages, apartment tower, and life-care home). Interviews were conducted at 2-year intervals. Residents were selected via systematic sampling procedures at each site.

A series of questions posed to subjects measured activity levels and a person's outlook on life. Within the latter set of questions, sub-areas of inquiry focused on: 1) age self-perception; 2) an emotional adjustment adjective check-list; 3) attitudes toward retirement; 4) morale; 5) if subject "found a life they were looking for"; and 6) whether they had moved between testing intervals.

Results were compared with interviews obtained from 600 matched control subjects living in dispersed housing. The data showed that in terms of check-list activities, clubs, and general social life, residence in two and in some instances four of the retirement housing settings were associated with more leisure activities than was residence in dispersed housing. The results concerning subjects' outlook on life were associated only moderately with the level of participation in leisure activities. A number of possible reasons were discussed regarding the lack of a strong statistical relationship. The majority of activities reported by subjects did not require special facilities in retirement housing. This type of housing contributed towards maintaining a relatively high activity level contrasted to the decline in activity level shown over the 2-year period by the matched dispersed housing controls.

Concerning the relationship between subjects' activity level and outlooks on life, it was concluded that because of different attitudes, style of aging, and various motivational factors, retirement sites should provide for persons who do, and in turn, for those who choose not to participate in leisure activities. For persons who do wish to participate, the environment needs to support such behaviors to the maximum extent.

"Mutual Assistance and Support in Retirement Housing," _Journal of Geron-tology,_ Vol. 31, No. 4, July, 1975, pp. 479-483.

KEY WORDS:

Man	Setting	Relational Concepts	Outcomes
Elderly	_Six retirement housing sites_	_Mutual assistance support networks; relocation_	_Patterns of inter-dependence_

This study focused on the networks of mutual assistance that develop among neighbors in age-concentrated housing for the elderly. The
extent to which these networks serve as functions that compensate or
replace mutual assistance with children has been debated in previous
studies. Three hypotheses were tested: 1) due to elderly residents'
decrease in contact with children because of controls in housing, there
will be less mutual assistance with children; 2) since there is
generally more contact with neighbors in this type of elderly housing,
there will be an increase in mutual assistance with neighbors; and 3)
there is a compensatory relationship on an individual basis between
mutual assistance with children and mutual assistance with neighbors,
especially in instances where residents receive little or no help from
their children. Six hundred interviews with individual residents were
conducted in six facilities for the well-elderly with types of
assistance reported by matched controls living in dispersed housing.
Controls were matched separately for each site according to ten
demographic variables. An earlier paper by this author reported that
site residents, relative to their environmental controls, had less frequent contact with children but more frequent contact with neighbors.

Results of the first hypothesis indicated that although test residents did live farther from their children and had less overall contact,
they were not being neglected relative to controls in terms of help
received from children. For hypothesis two it was found that there were
no clear, distinct patterns of differences between site residents and
controls with respect to mutual assistance patterns with neighbors. The
pattern varied from site to site. For the third hypothesis regarding
retirement housing, residents did not suffer from an overall lack of
assistance from children, nor did these residents benefit from greater
assistance from neighbors. This occurred despite large test-control
differences in interaction patterns between the six elderly housing
sites.

The consistent patterns of resident interdependency are not
explained solely on the basis on persons' health needs and abilities.
The author concluded that "retirement housing residents maintain the
style of interdependence they have built up through many years."
Residents did not try to "fill in gaps" in assistance they received
before they moved into retirement housing. Also, those who were
accustomed to greater autonomy and independence prior to moving in had
no need or desire to change their behaviors. More mutual assistance
with neighbors was found at two sites, less assistance at two sites, and
no differences were detected at two other housing sites.

"Seating Position and Interaction in Triads: A Field Study," *Sociometry,* Vol. 39, No. 2, 1976, pp. 166-170.

KEY WORDS:

Man	Setting	Relational Concepts	Outcomes
Human	*Tables*	*Seating position*	*Social interaction*

Research studies focusing on ways in which people interact within a
particular seating arrangement have shown a correlation between
proximity, placement and the amount of social interaction. An hypothesis was tested using real work observations and mixed-sex triad groups.

Forty-eight psychology students observed 201 naturally occuring
social triads at cafeteria tables. Each group was observed for approximately 20 minutes and observations were recorded concerning: 1) length
of conversations; 2) frequency of speech by each member of the triad;
and 3) lengths of acquaintances between members following observer
identification. Each observer was asked to summarize his findings into
a hypothesis that would explain which of the three members would speak
the most and which dyad of members would have the most social
interaction. Since each group was necessarily composed of one member
sitting alone on one side of a table and facing the other two who were
seated side-by-side, analysis in a three-way regression would give
insight into the relative effect of seating patterns on social behavior.

In conformity to past hypotheses, the authors found that a greater
amount of social interaction occurred between the dyad of the two members seating directly opposite one another. Since approximately half (N
=112) of the observations of this experiment were made with mixed-sex
groups, no correlation was found between sex and group interaction.
Neither did lengths of acquaintances have an effect on the amount of
talkativeness. In general, the authors concluded that the most important factor in the amount and degree of triad interplay was the ecological variable of seating position. This result tended to reinforce past
studies which suggested that proximity and arrangement in space are the
prime motivation factors of social intercourse.

"Visual Exposure, Form Uniqueness, Social and Functional Significance:
Towards a Predictive Model of Image Formation," *Man-Environment
Systems,* Vol. 7, No. 6, November, 1977, pp. 321-331.

KEY WORDS:

Man	Setting	Relational Concepts	Outcomes
Residents	*Urban area*	*Exposure; uniqueness; significance*	*Predictive model; landmark imageabi-lity*

Persons' recall of the physical environment of their city and how these persons' perceptions and cognitions are affected by the attributes of the actual setting were the focus of concern. Effects of visual exposure, attention demandingness and the social and functional significance of path and landmark elements were analyzed via data collected from a group of 429 urban residents. The authors stated that a need exists for planners and designers to better understand how the physical setting itself influences the representation of cognitive thought processes so that cities might be designed to possess elements that evoke more meaningful and comprehensive images to their users. Five hypotheses were developed and tested by the authors.

The research design was divided into three parts: 1) data utilized from a city image study done in 1974 in Columbus, Ohio; 2) a field study involving the collection of element exposure data; and 3) the measurement of three landmark attributes: form significance, social significance, and use significance, by having a panel of judges rank photographs of 16 local landmarks. In part one, data were collected by interviewers using a structured, open-ended questionnaire. In part two, landmarks and paths recalled by 5% of subjects were utilized. In part three, judges ranked the 16 photographs according to three criteria.

Subjects' responses to the three parts of the study indicated that: 1) exposure of landmarks as defined in this study (exposure=potential time in view X area visible X traffic volume) was not found to be a significant factor affecting their recall; 2) potential time in view of the 16 landmarks exhibited a very low positive correlation with recall; 3) a significant relationship existed between how much of a landmark was visible and its recall ($r = 0.82$, $p<.05$); 4) the number of people driving within potential view of a landmark was strongly correlated with its recall; 5) path exposure was significant for recall ($r = 0.54$, $p< .05$); 6) form uniqueness and social significance of landmarks was correlated with recall; 7) a strong relationship existed between the six variables when collapsed and the landmark recall frequency; and 8) an equation was formulated for predictive purposes. A number of study limitations and directions for future research were outlined.

The results suggested that if socially significant buildings were in highly visible locations, these structures would possess strong imageability. Form uniqueness was also seen as contributing to a building's imageability. A number of design and planning implications of the results were discussed. It was concluded that a more memorable and useable urban environment is attainable if the dimensions of landmark and path imageability can be incorporated into the built environment.

"Results of Investigations of the Microclimate in Some New Apartment
Buildings in Wroclaw" *Building Science,* Vol. 7, No. 1, March, 1972,
pp. 18-22.

KEY WORDS:

Man	Setting	Relational Concepts	Outcomes
Inhabitants	Apartments	Microclimate, environment, thermal	Comfort criteria revision

This investigation examined the air temperature and relative humidity of one hundred apartments located in Wroclaw, Poland during the
years 1964-1966. Also considered were the influences of local climate
conditions and housekeeping practices on the microenvironment. The aim
of the investigation was thus to define the average parameters of the
microclimate variables examined, and to determine the ways in which the
local climate, cooking, washing, and window ventilation affects these
variables.

The 100 apartments in the sample were located in 24 buildings
of various types situated throughout Wroclaw. Each apartment was
required to fill out a detailed questionnaire concerning: 1) the number
of inhabitants; 2) location of particular rooms; and 3) their
utilization. Climatic conditions of the three years during which the
study took place were obtained from the Department and Observatory of
Meteorology and Climatology of the University of Wroclaw. Air temperature and relative humidity readings were gathered daily for: 1)
sitting rooms, 2) bedrooms, 3) kitchens, and 4) bathrooms.

During the summer season, May to December, the air temperature
ranged from 18.5°C to 23°C, with a relative humidity of 61-74 percent.
In the heating season the air temperature and relative humidity were
lower (17.5°- 19.5°C and 46-68 percent RH). It was discovered that
during the summer months local climatic conditions produce a dominating
influence on the microclimate, with housekeeping functions and routines
being practically negligible. In contrast, due to the influence of
central heating, the climatic conditions during the winter months did
not tend to be as much of a factor, but such activities as cooking did
become a factor. Taking baths and washing in the bathroom tended to
affect the microclimate of the rest of the apartment only slightly,
whereas clothes washing and drying had a marked influence on the relative humidity, and to a lesser degree, the air temperature. Thus, the
author concluded that "the lack of officially established boundary values for the particular microclimate element seriously impedes the
achievement of a satisfactory comfort level of the microclimate in
dwellings."

"Intraurban Location of the Elderly," _Journal of Gerontology_, Vol. 30, No. 4, July, 1975, pp. 473-478.

KEY WORDS:

Man	Setting	Relational Concepts	Outcomes
Elderly	_Residences_	_Intraurban location_	_Distribution_

This study examined the influence of a set of factors on the intraurban location patterns of the elderly population of an urban area. The majority of prior studies in this area of research have concluded that the aged are concentrated in urban cores as opposed to being equally distributed throughout urban areas. Correlation analysis was utilized by census tracts in Toledo, Ohio to measure the relationship between the residential location of the elderly and these variables: 1) distance to the Central Business District; 2) age of persons' neighborhood, 3) median value of housing; 4) the numbers of multiple-family dwelling units; and 5) percentage of total non-elderly in the area. The spatial pattern of the elderly differed from that of the non-elderly in this urban area. Three analyses were performed for the census time period.

Relationships between five of the six variables were statistically significant, although these links were weak. At no time between 1940 and 1970 were more than 40% of the variations in the elderly population distributions explained. The three analyses did not support the generalizations found in the literature concerning the location of the elderly.

It was concluded that more research is needed to understand elderly persons' residential patterns. Problems with methods of collecting and analyzing data need to be resolved. Census tract data was viewed as often vague and misleading. Micro samples of data elicited from interviews can also misrepresent and distort the focus of the research. Improved methods of collecting data on this population group were encouraged by the authors so that planning and design decision making may become more responsive to the actual needs of the elderly in urban areas.

"The Effect of Relocation on the Satisfaction of Psychiatric Inpatients," _Journal of Clinical Psychology,_ Vol. 32, No. 4, October, 1976, pp. 845-848.

KEY WORDS:

Man	Setting	Relational Concepts	Outcomes
Psychiatric inpatients	_Hospital_	_Involuntary relocation_	_Adjustments, satisfaction_

Fifty-one psychiatric inpatients were tested on a 36-item satisfaction scale before and after a reorganization at Harlem Valley Psychiatric Center. Literature was reviewed which addressed involuntary relocation and increased mortality rates, increased levels of plasma cortisol, increased pessimism regarding health outlook, attitudes toward life, and decreased levels of psychosocial activity. The results of the bulk of these studies have indicated that the magnitude of such changes were often dependent on the extent of environmental change involved. On the other hand, results of studies with institutionalized psychiatric populations regarding involuntary relocation indicate no significant before - after changes.

In this study, patients were divided into three groups (low, medium, and high stress), on the basis of whether they experienced no change as a result of: 1) the social reorganization of the facility, 2) environmental change only (different staff, and different fellow patients). Of the total sample of 200 tested, 19 were in the high stress group. Nineteen patients each from the medium and low stress groups were selected for comparison with the high stress group. The samples were composed of adult (mean age, 47), chronic patients (mean length of hospitalization ranged from 133 to 138 months). Retesting of these patients (57 total) was started on the sixth day after the reorganization and continued for one week. Patients in the three groups were matched according to age, length of hospitalization, and sex. Contents of the 36-item satisfaction scale were determined on the basis of pilot results using a 74-item scale. Within each content area (general, ward, staff, treatment plan, activities, and food) 6 yes-no questions comprised the final testing instrument. The total satisfaction score was the number of items that were answered in the satisfied manner. A series of one-way ANOVA tests were utilized to analyze data.

Results indicated that there were no reliable differences among the satisfaction scores of the groups as a result of the relocation to a new spatial environment. This supported the hypothesis that the involuntary relocation of chronic psychiatric patients to another institutional environment results in few noticeable effects. The failure to find significant overall differences is explainable in that relocation resulted in decreased satisfaction levels in some patients and increased levels in other patients which nullified each other in the pretest - post test group comparisons. Based on the findings of this and prior studies by others, the authors concluded that chronic mental patients have the ability to adapt with relative ease to dramatic environmental changes, and that this might occur because many of these patients tend to be socially and environmentally withdrawn either as a result of their illness or due to the effects of prolonged institutionalization.

SMYLIE, H.G., A.I.G. DAVIDSON, A. MACDONALD AND G. SMITH   227

"Ward Design in Relation to Postoperative Wound Infection: Part I"
British Medical Journal, Vol. 1, January 9, 1971, pp. 67-72.

KEY WORDS:

Man	Setting	Relational Concepts	Outcomes
Injured patients	Hospital wards	Treatment setting design	Cross-infection control

In order to control postoperative wound cross-infection a ward was
designed which controlled ventilation and provided as many separate
rooms as was feasible. Based on the assumption that a higher percentage
of efficiently ventilated single-bed rooms would minimize cross-infection, a four-year study compared the results of open ward treatment to
the new design.

During the first two years of the study patients resided in an open
ward environment; whereas in the final two years patients were placed in
the new ward design. Staff administration, observational methods, and
the analysis of results remained constant throughout the four-year
period. Care was taken to maintain the same surgical habits throughout
the study. The salient difference in ward architecture was the division
of ward space into separate compartments so that controlled ventilation
was more possible. Air was circulated in the new environment by a slow
continuous exchange system resulting in a rate less than two air changes
per hour.

The results revealed that following transfer, postoperative wound
infection was reduced by approximately 55%. Through examination of certain types of infection it was shown that the transfer was followed by a
72% reduction in cross-infection type wounds. One particular strain
virtually disappeared in the new environment. The authors point out
that many months after the disappearance of the epidemic strain from the
new ward, it was again detected in an open surgical ward in the same
hospital. The authors concluded that the control of cross-infection is
attainable through the proper design of the surgical ward.

SNYDER, R. AND WILLIAMS, A.   228

"The Locating Choices of Studies in Lodging and Flats," *Urban Studies*,
Vol. 10, No. 1, February, 1973, pp. 87-90.

KEY WORDS:

Man	Setting	Relational Concepts	Outcomes
Students	*University*	*Transportation, housing*	*Public transporta-tion, needs for*

Using the up-to-date voter registration files in York, England, the
number of students per 100 voters was determined for each of 50 polling
districts in the year 1972. These students were then interviewed to
identify how they made their choice of living quarters.

Many variables were analyzed in this study with no significant differences noted in such factors as the number of graduates living in
flats as opposed to undergraduates living in single rooms in homes
offering lodging. The most significant variables in determining where
the students lived proved to be cost of travel to the university.
Student density was greatest in the sections of the city which were most
accessible to the university. This meant that those areas which could
be reached on foot were densely populated, but because close-in lodgings
were more expensive, the greatest density of students' living quarters
were in locations which were more accessible to the university by bus.

As a result of this study, a recommendation was made to both the university administration and the city council to extend one particular bus route one mile further. This change would produce approximately 50-60 additional places for students living in flats or homes.

One important implication for university policy would be that improvements in the public transportation system should be substituted for at least a part of the university's building program involving housing, since more students could be accomodated at less cost through this approach.

SPINETTA, J.J., D. RIGLER, AND M. KARON   229

"Personal Space as a Measure of a Dying Child's Sense of Isolation," *Journal of Consulting and Clinical Psychology*, Vol. 42, No. 6, December, 1974, pp. 751-756.

KEY WORDS:

Man	Setting	Relational Concepts	Outcomes
Children	*Hospital*	*Personal space*	*Preferred psychosocial distances*

Interpersonal distance measures were used in an attempt to identify the sense of isolation present in a child who possessed a fatal illness as he or she neared death. Literature was reviewed that measured a person's space through the use of the psychosocial distance concept. This enabled the researchers to determine whether and how much an individual wished to associate with, or disassociate from others. The present study was grounded in the conceptualization of the hospitl environment as a complex set of cues that enables the reseacher to elicit information concerning the ill child and significant adult figures in the child's hospital life.

Two hypotheses were formed: 1) that children aged 6-10 years with a fatal prognosis would place scale figures representing meaningful adults in their hospital environment at a greater distance than would children who do not have a fatal prognosis; and 2) that children aged 6-10 years in subsequent admissions to the hospital would place the figures at a greater distance than would those admitted to the hospital for the first time for the same illness.

A battery of tests were administered to 50 children hospitalized in Children's Hospital in Los Angeles. A scale model replica of a typical hospital room at the hospital was designed at a scale of 1 inch to 1 foot, with all furnishings and human figures accurately scaled. Each child was told a series of hypothetical stories involving each of four adult figures (nurse, doctor, mother, father) and asked to place the figures in the room accordingly. No time limit was set for this task. Of the 50 children, 25 were hospitalized with a diagnosis of leukemia, while the other 25 had chronic but non-fatal illnesses. Members of the two groups were matched as closely as possible according to a number of variables.

The results showed that for the figure placement task the group of 25 leukemia children placed the figures at a distance significantly greater than did the matched control group of chronically ill hospitalized children. Although the distance of placement increased with both groups in subsequent admissions, the leukemic children increased the distance significantly more than did the chronically ill. The author believed that this lent strong support to the hypothesis that the sense of isolation grew stronger as the child neared death.

A final placement of preferred figure distance led to the conclusion that the 6-10 year old fatally ill child not only perceived a growing psychological distance from those around him or her, but also preferred it that way. The built environment was referred to as one of many possible influences on these results. The authors placed great weight in the explanation of these results on children's social learning theory processes.

STOKOLS, D, R.W. NOVACO, J. STOKOLS, AND J. CAMPBELL   230

"Traffic Congestion, Type A Behavior, and Stress," _Journal of Applied Psychology,_ Vol. 63, No. 4, August, 1978, pp. 467-480.

KEY WORDS:

Man	Setting	Relational Concepts	Outcomes
Commuters	Simulated highway trips	Stress judgements, trip type	Impedance levels, physiological consequences

A simulated environmental study was conducted to assess the effects of routine exposure to traffic congestion on the mood, physiology, and task performance of automobile commuters. Literature was reviewed that focused on the urban environment and its impact on people, and how this sizable body of research has looked at the behavioral and health consequences of exposure to numerous urban settings, including high-rise buildings, crowded markets and department stores, and noisy neighborhoods.

The many effects and after-effects of commuting and traffic congestion often have a profound impact of the individual. The major goals of this study were: 1) to develop a conceptualization of traffic congestion and its impact on people over extended periods, 2) to establish measurement criteria for operationalizing this conceptual framework; and 3) to test certain hypotheses derived from the framework (within the context) of the field experiment. Traffic congestion was conceptualized as an environmental stressor that impedes one's movement between two or more points. The degree of impedance encountered by travelers was simulated according to: 1) the distance traveled between origin and destination, and 2) the amount of time spent in transit between these points. Two major hypotheses were developed and tested based on the above discussions.

Sixty-one male and 39 female industrial employees were assigned to low, medium, or high impedance groups on the basis of the distance and duration of their commute and were classified as either Type A or Type B on a measure of coronary-prone behavior. The degree of congestion was assessed by means of a 9-point bipolar scale included in the daily travel log. Physiological arousal was measured in terms of systolic blood pressure, diastolic blood pressure, and heart rate. Task performance was measured by: 1) the number of attempts made by each subject on Feather's (1961) insoluble puzzles; 2) the number of boxes corrently completed on the Digit Symbol Task; and 3) the number of symbols recalled in this task. Subjects' moods were assessed via nine 5-point Likert scales. Also, a series of background questions were administered to each subject. Pearson correlations between these variables were computed, a series of multivariate analysis of variance were performed on major dependent variables and 2-way ANOVA tests were performed on the background data which was collected.

The results showed that subjective reports of traffic congestion and annoyance were greater among high and medium impedance commuters than among low-impedance subjects. Also, commuting distance, commuting time, travel speed, and number of months enroute were significantly correlated with systolic and diastolic blood pressure. Contrary to the authors' prediction, medium impedance A's and high impedance B's exhibited the highest levels of systolic blood pressure and the lowest levels of frustration tolerance among all experimental groups. The results were discussed in terms of the degree of congruity between commuters' expectations and experiences regarding travel constraints.

STOKOLS, D. W. OHLIG, AND S.M. RESNICK   231

"Perception of Residential Crowding, Classroom Experiences, and Student Health," *Human Ecology,* Vol. 6, No. 3, 1978, pp. 233-252.

KEY WORDS:

Man	Setting	Relational Concepts	Outcomes
College students	*Residences, classrooms*	*Crowding*	*Health*

It was proposed by the authors that a typology of crowding experiences based upon physical as well as social dimensions could be used to predict a person's sensitivity to specific environmental situations. Specifically, this study attempts to show a correlation between a subject's perceived level of residential and classroom crowding and his or her subsequent need for health care.

Twenty-seven undergraduate students within a large lecture course were chosen statistically for their participation in this study. The subjects lived in suite-type dormitory rooms or similar off-campus housing. Semantic differential questionnaires were administered to the students to measure their perception of specific residential and class indices of crowding. The experiment was designed to elicit attitudes that the subjects experienced and to compare these attitudes with health and academic data. Correlations between residential and classroom assessments, regression analyses between perceptions and health data, and multivariate analyses of variances between residential evaluations were used to provide empirical indices for the study.

Results of the various analyses indicated that the subjects' evaluations between residential and classroom crowding showed a high correlation (r = .6). Likewise, the evaluations concerning residential
crowding were directly related to frequency and type of medical attention during the year (r =.716). "Perceived residential crowding and
negative perceptions of residential social climate were strongly associated with increased sensitivity to crowding in a classroom situation,
impaired course performance, and visits to the student health center."

STORANDT, M. I. WITTELS, AND J. BOTWINICK   232

"Predictors of a Dimension of Well-Being in the Relocated Health Aged,"
Journal of Gerontology, Vol. 30, No. 1, January, 1975, pp. 97-102.

KEY WORDS:

Man	Setting	Relational Concepts	Outcomes
Elderly	_Residential_	_Dimension of well-being, relocation_	_Successful adjust-ment, predictors_

    The study examined a number of measures in response to the need to
identify whether poor adjustments of the elderly to new residential
environments are indicative of the actual setting or responses to
(transposed upon) the residential environment. These measures were examined to determine dimensions useful as screening devices for elderly
persons applying to housing facilities that required independent living
and self-care. Test performances of elderly persons were monitored from
the time of their move into an apartment complex where independent
living - self-care was necessary. After living in their apartments
approximately 15 months, subjects were rated for how well they were
faring in respect to a clinical dimension ranging from "played out"
(senile) to "vital" (healthy).

    One hundred twenty two subjects aged 61 to 88 years were given a
wide variety of tests. The majority of apartments in the new environment were efficiencies while the remainder were one bedroom units for
couples. Each resident had to meet all federally subsidized housing
financial-health qualifications. The series of tests were administered
to subjects individually at pre-determined time intervals. The testing
session lasted two hours. The types of tests were purposely selected to
relate to issues most prevalent in the literature. This battery of
tests included four groupings of procedures plus an array of demographic
functioning; 2) personality and morale; 3) health status and health-
related habits; and 4) activity levels and past times.

    Test scores were analyzed in relation to these four rating cate-
gories plus the demographic data. A multivariate analysis of variance
demonstrated significant differences (f = 1.61, df = 100/371.13, p<
.001), which then permitted a series of univariate analyses. A discriminate function analysis was conducted to determine those test procedures most predictive of the clinical rating categories.

Of the twenty-five predictor variables, 14 were related to the
descriptive dimension of well-being. Two were especially prominent -
the WAIS Comprehension subtest and the Crossing-Off psychomotor task.

Although the authors did not endorse this dimension as a sole criterion of subject assessment of new living environments, they felt that
the rating of well-being was important within the total scope of
research concerning the elderly. The predictors of poor adjustment to
relocation were not viable in this sample of healthy elderly persons.
It was concluded that for healthy persons aged 62 and over, results of
this study indicated that neither age, sex, nor marital status should
deter them from changing place of residence. These factors were not
found to be significant in the prediction of these elderly persons'
well-being approximately 15 months after relocation.

STORANDT, M. AND I. WITTELS   233

"Maintenance of Function in Relocation of Community-Dwelling Older
Adults," *Journal of Gerontology,* Vol. 30, No. 5, September, 1975, pp.
608-612.

KEY WORDS:

Man	Setting	Relational Concepts	Outcomes
Elderly	*Institutional/ non-institutional residential setting*	*Relocation*	*Adjustment*

An examination of behavioral test performance was conducted in a
pre-post test design, comparing four groups of relatively healthy
elderly persons in the examination of non-movers as opposed to voluntary
movers. Recent studies were cited concluding that the relocation of
institutionalized elderly populations can have a life-endangering effect
on these persons. The high-risk stress of moving to a new environment
has rarely caused detrimental health states when healthy persons were
the subjects of investigation. The death rate after relocation was not
studied as a primary issue in the present experiment.

The four groups were: 1) movers-building 1, which consisted of 39
women and 12 men who were relocated from independent community dwellings
to a new high-rise apartment complex for the elderly which was located
nearby; 2) non-movers building 1, consisting of 22 women and 6 men.
These persons chose not to move for a variety of personal reasons.
Other applicants were either on a waiting list for the new complex or
were disqualified on the basis of being judged as incapable of independent living; 3) the movers-building 2 group; consisting of 31 women and
7 men who were primarily native born (St.Louis) as opposed to the
largely Jewish population of group 1; and 4) the non-movers building 2
group; consisting of 5 women and 1 man who were applicants to building
2, who for personal reasons chose not to move into the complex. All
subjects were tested prior to the opening of the new complex and one
year later. Demographic characteristics of the four groups were analyzed (age, years of education, socio-economic status, and a physician's
medical rating of each individual). All groups were tested for com-
patibility prior to both moving or non-moving conditions.

Pre-tests were given (for movers) from 5 months prior to relocation. An assessment battery was employed containing 21 procedures.
Four areas of function were measured: 1) cognitive and psychomotor
performance; 2) personality and morale; 3) health, and 4) activity
patterns.

The results revealed no decrement in function among those who
changed residence. A series of statistical tests measured the effects
of the four conditions for the contents of the testing battery. The
three stages of analysis for both the movers and non-movers groups
corresponded to: 1) initial testing; 2) a follow-up phase; and 3) the
measure of change.

The fact that the sample groups were screened for health status,
relocated voluntarily to a new environment in the same area, and faced
with improved social and living conditions was discussed as one possible
factor in their successful adjustment after relocation. Also, volunteers comprised the sample. This might have biased the data since these
sub-groups of residents in the complexes might have been the most
"capable" of the total populations of the four groups. Many long-
standing habits and customs were uninterrupted with the move.

It was concluded that the arguments against relocation of older
adults are best tempered and understood following a careful examination
of the conditions surrounding relocation of institutionalized aged.
This study has shown that there is no necessary comparable risk in relocation among relatively healthy, well-adjusted elderly persons who do
not reside in institutional environments.

STUTZ, F.P.   234

"Adjustment and Mobility of Elderly Poor Amid Downtown Renewal,"
Geographical Review, Vol. 66, No. 4, October, 1976, pp. 391–400.

KEY WORDS:

Man	Setting	Relational Concepts	Outcomes
Elderly males	*Hotel*	*Life support networks*	*Disruption*

A major consequence of urban renewal programs is the dislocation of
relatively stable but politically isolated groups or individuals. One
such group is the elderly poor who occupy the low cost hotels of most
American urban centers. This study attempts to accumulate a certain
amount of data about one such population and present preliminary policy
recommendations when this group is faced with environmental change.

Thirty-two male residents of the Golden West Hotel in downtown San
Diego were interviewed from a total population of 400. The median age
of a resident was 70 and most of the occupants tended to be stable, sub-
sisting on social security or retirement insurance. Unlike some
surrounding occupants of Salavation Army missions or welfare recipients,
these residents considered themselves independent and self-sufficient;
the hotel being a "home" rather than a "haven".

The major problem faced by the group was the limited degree of mobility that accompanies old age and subsistence level incomes. The study showed that most of the essential services of the residents were confined to a two-square block area surrounding the hotel. Excursions of more than four or five blocks by bus were viewed as luxury outings. Renewal plans being considered by the city would drastically curtail the residents' access to food and health services and cause disruptions in public transportation services.

The author concluded that eldely residents of downtown hotels form a major part of the urban environment. Mobility to essential services in renewal efforts must be insured to downtown residents and public transportation should be molded about the needs of the elderly poor. Any attempt to alter a downtown urban environment should be based on human as well as economic physical considerations.

TAINSH, M.A.   235

"The Symptoms Reported by Long Distance Travellers," *Applied Ergonomics,* Vol. 6, No. 4, December, 1975, pp. 209-212.

KEY WORDS:

Man	Setting	Relational Concepts	Outcomes
Travellers	*Car, train*	*Travel frequency*	*Health status*

Frequency of traveling long distances for business purposes was seen as potentially increasing the number of physical ailments of the travelers. Specifically, the author hypothesized that: 1) health indices would decrease with increased car and train travel; 2) age would be a significant factor in the frequency of health problems; and 3) the personality of the traveller would have a marked effect on the rate of health complaints.

One-hundred and sixty-five white collar workers from sixty English organizations participated in the study over a period of two years. Only those business trips over 100 miles or more were studied in order to isolate the effects of long distance travel as opposed to commuting type travel. The subjects were interviewed at their place of work after completing the travel and were questioned about specific health problems as well as being administered the Eysenck Personality Inventory. Health indices were measured by such traits as headaches, backaches, anxiety, and fatigue. Originally, fourteen indices of health were categoried: ten of which were found to be statistically significant.

Results of the surveys showed that train travel over extended distances resulted in eight significant health problems, while traveling by car resulted in only one malaise. Frequencies of health problems rose rapidly as increased train travel occurred, while complaints associated with car travel remained relatively constant. It was also found that personality factors were related significantly to car travel but not to train travel. Contrary to Eysenck's theory, the studies showed that personality is important in determining symptoms associated in car travel. Finally, the author concluded that age and economic position

were of little relevance in determining health problems following either
train or car travel.

TAYLOR, L.H. AND E.W. SOCOV   236

"The Movement of People Toward Lights," *Journal of the Illuminating
Engineering Society,* Vol. 3, No. 3, April, 1974, pp. 237-241.

KEY WORDS:

Man	Setting	Relational Concepts	Outcomes
Human	*Laboratory*	*Lighting levels*	*Mobility prefer-ences*

The purpose of this study was to determine in which directions
people would move when presented with choices of lighting levels in different avenues of movement. The authors hypothesized that people, given
an equal lighting level in two paths, will: 1) more often choose the
right path, given unequally lit paths; and 2) will choose the brighter
of the two. Also an illumination ratio was hypothesized as existing,
influencing behavior pattern changes.

One hundred eleven male and female adults participated in the
experiment. Subjects ranged in age from 18 to over 50 years. Each subject was admitted to the center of a twenty foot corridor and presented
with a sign that was designed to take their attention from the lighting
levels at the ends of the corridor. One end of the corridor was lighted
at a constant level of 1 footcandle (fc) while the opposite end was
illuminated at 1, 3, 10, 30, or 100 fc. Both the constant and variable
ends were switched from left to right, making ten different cominations
of lighting variables possible. An observer recorded which direction
each subject proceeded after reading the sign and which way he returned
from the experimental setting.

Sixty seven percent of the subjects chose the right path when presented with a 1:1 illumination ratio. This choice was in support of the
authors' original hypothesis. It was also found that 70 percent of the
subjects chose the brighter path regardless of direction. The data
indicated that directional behavior was reinforced as the ratio
increased and the most dramatic preference distinctions were recorded at
between 1:10 and 1:30 ratios. "A behavioral change threshold exists
between illumination ratios of 10 and 30. Below the threshold the
leaving results closely duplicate the entering results but above this
threshold the entering and leaving results exhibit large deviations."

TAYLOR, L.H., E.W. SOCOV, AND D.H. SHAFFER   237

"Display Lighting Preferences," *Journal of the Illuminating Engineering
Society,* Vol. 3, No. 3, April, 1974, pp. 242-248.

KEY WORDS:

Man	Setting	Relational Concepts	Outcomes
Human	*Laboratory*	*Lighting*	*Preferences*

In order to establish a number of basic lighting rules for the
display of objects, the authors have performed a series of preference
experiments with a variety of lighting arrangements and illumination
levels. It was hypothesized that six different criteria for judging the
display could be broken into subsets of independent lighting theorems.

One hundred thirty-five male and female subjects were used in this
experiment and were chosen across a wide band of demographic factors.
Each student was shown two displays of an object simultaneously lighted
by a randomly chosen arrangement of six possible configurations. As the
subject was shown each set of objects he was asked: 1) which attracted
his attention first, 2) his preference, 3) which is better displayed, 4)
the more attractive, 5) the more dramatic, and 6) was asked to walk to
the box which appeared to draw him more. Each subject observed four
separate display sets and asked the above six questions each time. Sets
were randomized and statistical tests showed no variances with respect
to left or right preferences.

The results of the data indicated that the original six questions
could be reduced to three independent questions concerning how attrac-
tively the object is lighted, how much attraction the lighting induces,
and how dramatic the lighting effect is. It was found that front spot
lighting is preferred when secondary lights are used. Top lighting and
back lighting came next in preference order. Incandescent (or spot)
lighting is preferred over floursecent (or area) lighting, gives better
displays, and yields more attractive objects. Lighting intensity was
found to have a relatively minor effect on the preference structure and
was only effective in initially attracting the subject's attention.

TEAFF, J.D., M.P. LAWTON, L. NAHEMOW AND D. CARLSON   238

"Impact of Age Integration on the Well-Being of Elderly Tenants in
Public Housing," *Journal of Gerontology,* Vol. 33, No. 1, January, 1978,
pp. 126-133.

KEY WORDS:

Man	Setting	Relational Concepts	Outcomes
Elderly	*Public housing*	*Age integration and well-being*	*Positive impact, age-segregation*

This report addressed the question of whether housing facilities
for the elderly should be age-segregated or age-integrated. The
integration of the elderly into the mainstream of society needs to be of
primary concern to gerontologists, social planners, and environmental
designers. The literature on these aspects of housing largely supports
the view that age-segregation is beneficial when tempered with a certain
degree of interaction with persons representing other age groups.

The social, psychological, and environmental situations of a
national area probability sample of 1,875 elderly tenants in 153 public
housing sites were studied by means of interviews with tenants and project managers. Additional data was collected via direct observation of
an array of social factors and environmental attributes of the housing

facilities. The emphasis was on the association between the age mix and well-being of residents in the housing facilities. Housing sites were primarily located in urban areas. Interviews were conducted with subjects on an individual basis.

Indices of well-being were: 1) activity participation, 2) family contact, 3) morale, 4) housing satisfaction, 5) mobility, and 6) friendships. Control variables were selected from: 1) persons' demographic characteristics, 2) the predominant social context of tenants living in a housing project, and 3) physical characteristics of the housing environment. Multiple regression analysis was utilized in the analysis stage. Housing site types were divided into six categories. These were: 1) random, 2) clustered, 3) mixed, 4) segregated within (high rise, low rise), 5) segregated-adjacent, and 6) totally segregated. Results indicated that for all control variables, age segregaton showed small, yet reliable relationships to housing site activity participation, morale, housing satisfaction, and neighborhood mobility. Depending on the particular measure of well-being, age-grading was utilized to account for specific findings. This was observable through persons' increased stress and behavioral restrictions imposed by environmental constraints, i.e. personal security and proximity to disruptive factors.

In conclusion, age-segregated living was associated with positive outcomes in these public housing sites. It was shown that options available to the individual must be maximized. Both social policy planning and the physical environment must accommodate the need for elderly persons' freedom of choice.

TENNIS, G.H. AND J.M. DOBBS   239

"Sex, Setting and Personal Space: First Grade Through College," *Sociometry,* Vol. 38, No. 2, September, 1975, pp. 385-394.

KEY WORDS:

Man	Setting	Relational Concepts	Outcomes
Students	*Urban schools*	*Age/sex differences, interpersonal distance*	*Age-related interpersonal distance*

This study examined the effects of age, sex, and setting upon interpersonal distancing preferences among 1st, 5th, 9th and 12th grade white public school students and college sophomores. Subjects were tested two at a time in same-sex pairs in both "corner" and "center" of a designated area. Direct measures of distancing were analyzed and showed interpersonal distance to be greater among older than younger children, greater between males than females, and greater in the corner than in the center setting.

Sex differences were noted more among older than among younger subjects and the youngest children maintained closer interpersonal distance in the corner setting while all other subjects were closer while holding conversations in the center setting. Interpersonal distance preferences seem to indicate that older males are aware of society's view toward physical closeness between adult males. The 1st grade male has not become aware of this cultural viewpoint and the 1st grade male maintained even closer distance than did the 1st grade female subjects.

Somewhere between 1st and 5th grade a transition occurs wherein more distance in the "corner" setting reflects more fear of being "cornered" - of having one's escape route cut off. The youngest children were the most trusting of each other in the corner setting.

The use of direct measuring of interpersonal distancing, rather than relying on information obtained through questionnaires or interviews, seems to reflect a truer picture of distancing patterns among younger persons.

The findings of this study provided insight into the development of personal space boundaries of the individual through the maturation process.

TOGNOLI, J.   240

"The Effect of Windowless Rooms and Unembellished Surroundings on Attitudes and Retention," *Environment and Behavior,* Vol. 5, No. 2, June, 1973, pp. 191-201.

KEY WORDS:

Man	Setting	Relational Concepts	Outcomes
Students	Laboratory	Environmental setting	Behavior

The purpose of this study was to establish the relationship between the environmental quality of a room and the resultant behavior of people exposed to that environment. The author hypothesized that a room with an exterior view, furnished with comfortable chairs, and pleasantly decorated, will lead to a more positive human response than one without the above embellishments, but will also serve to provide a higher degree of distraction.

Fifty-six male and female undergraduates at C.W. Post College were given tests in rooms that were arranged in one of eight different configurations. The independent variables of the experiment were either a hard or soft chair, a room with or without an exterior view, and a room embellished with decoration or having only bare, painted walls. Seven subjects were tested in each of the eight combinational environments. The tests consisted of a standardized retention questionnaire used to measure the subjects perceived feelings of pleasantness, interest, distractedness, and comfort.

It was found that the combination of the three independent variables provided rather complex and unexpected relationships. Significant relationships were found between room embellishment and interest, pleasantness and presence of windows, and comfort and a three way combination of all three variables. Comfort, for example, was found

to be highest in conditions with soft chairs, no windows, and no embel-
embellishment or in rooms with hard chairs, decoration, and no exterior
view. In the retention test, it was found that the most advantageous
environments were those with soft chairs and windows or with hard chairs
and no windows. In a three-way analysis of variables concerning retention, the results mirrored those for the comfort rating. The author
concluded that the results of this study could not be generalized since
the relationships between the three variables were too complex and did
not fit the expected patterns predicted in the original hypothesis.

TUCKER, J.   241

"Population Density and Group Size," *American Journal of Sociology*,
Vol. 77, No. 4, January, 1972, pp. 742-749.

KEY WORDS:

Man	Setting	Relational Concepts	Outcomes
Students	*Rooms*	*Density inter-action*	*Greater density, decreased inter action*

    This study examined the relationship between the size of the small
interacting group and its environment. Specifically the study investigated the concept that as population density increases, the interaction
of a group will decrease proportionately.

    The major hypotheses generated by the author were: 1) persons
spending most of their time in more densely populated areas will gather
for social interaction in smaller groups than those who inhabit areas
which are more sparsely inhabited, 2) persons spending most of their
time in more densely populated areas will gather for social interaction
in fewer bisexual groups, and 3) as population density increases, males
tend to gather in smaller groups than will females. In other words,
there is an inverse relationship between population density and the
number of males in an interacting group as compared with the number of
females.

    The method used to collect data was to count the number of persons
in a group as they emerged from a cafeteria line and found their way to
a seating arrangement. The sample for this study was drawn from three
Texas university campuses with varying population densities.

    The findings supported the theory that as population density rises,
the interaction between persons decreases correspondingly. This is
explained as a psychological means of reducing the effects on the person
which result from increased population density. With the increase in
population density comes an increase in unavoidable intrusive interpersonal encounters which produces stress in a person. It was concluded
that if there are no practical means of physical retreat, the person
uses what means he has at his disposal to mentally reduce the interpersonal encounters. The result is less interaction with fewer persons.

Tyler, D.A.   242

"Noise and Truck Drivers," _American Industrial Hygiene Association_
Journal, August, 1973, pp. 345-349.

KEY WORDS:

		Relational	
Man	Setting	Concepts	Outcomes
Truck drivers	_Trucks_	_Design/noise_	_Noise reduction_

Truck drivers are often confined to the driving compartment for
extremely long periods of time. Here they may be subject to stresses
such as vibration, noise, air contaminants, heat stress, glare of lights
or sunshine, and other discomforting elements. This study examines the
exposure to noise drivers face while working.

The objectives were to identify: 1) the source and character of
the noise, 2) the noise reduction achieved, and 3) the noise reduction
techniques that were utilized to achieve the effect.

The studies of drivers' typical exposure in a vehicle with a four-
cycle diesel engine revealed that: 1) with all the windows and air
vents open, and while in heavy traffic, the noise level may be approximately 15 dBc higher than when all windows and vents are closed; 2) with
the windows and air vents closed, the engine operating noise predominated within the cab, 3) noise arising from air intake and exhaust
systems may add as much as 5 dBc within the cab, 4) the vibration of
loose parts exposure, or sympathetic panels may add a 3 dBc increase to
driver's exposure, 5) gear shifting noise may increase the noise level
temporarily by about 1 dBc, 6) the shutters opening and closing are
detectable and may account for a 1 dBc increase, and 7) tire noise
accounts for less than a dBc increase within the cab.

A sound level meter and an octave band analyzer were utilized for
measuring the noise levels. All vehicle windows and vents were fully
closed and all accessories were off except for the air conditioning
unit.

All testing was done on a typical driver's route used to make fuel
deliveries. Most routes included country and freeway driving.

The results of the study indicated that the noise levels in the
cabs could be adequately controlled with simple acoustical techniques
using common acoustical materials. This reduction could be achieved
with minimal cost and vehicle weight increase.

"An Exploratory Study of Patterns of Social Interaction, Organizatin,
and Facility Design in Three Nursing Homes," _International Journal of
Aging and Human Development,_ Vol. 4, No. 4, 1973, pp. 319-331.

KEY WORDS:

Man	Setting	Relational Concepts	Outcomes
Nursing home patients	_Nursing homes_	_Design/social interaction_	_Optimal interaction_

The basic question addressed by this study concerned relationships
between environmental design, management, the individual, and social
interaction, and what factors promote or deter social interaction.
Social interaction was examined according to three phases: 1) the tendency to congregate, 2) the ability to impersonally interact with
others, and 3) the capacity to converse. Since conversation has been
linked with rehabilitation, its promotion was stressed in the study
design.

Three extended care facilities were examined according to the
following criteria: 1) all facilities were to be certified as extended
care facilities receiving Medicare patients, 2) proximity to research
base of operations, 3) facilities which had had time to adjust to the
present management because the administrator would have had more time to
put his ideas into practice, 4) nursing homes with at least 100 beds, 5)
facilities of like age (mid to late 1960's) were selected, and 6) facilities of three types of sponsorship (private, voluntary, and public)
were sought.

The manner in which data were collected was through: 1) administrators' interviews, 2) descriptions of the physical environment,
including inventories of room furnishings and appointments, were systematically recorded, 3) photographs were taken in each area studied of
the room itself and of specific items of interest, 4) systematic observations were made of the residents as they occupied and interacted in
each social area, 5) guided interviews with 25% of the residents in each
facility, and 6) staff members were informally queried about their
observations on design functionality and resident use of various rooms.

Six field workers participated in a week long training period and a
series of observations and interviews at a home for the aged prior to
collecting data for the present study. The field teams made two site
visits to each of the three facilities. Exploratory purposes sought
general data rather than to concentrate on one facility in greater
detail.

The major findings of the study were as follows: 1) the title of a
room or designation of an area may have little bearing on the function
of that space; 2) socially designated spaces may be unused as settings
of social behavior for several reasons; a.) due to immobility, resident
were unable to either get to the dayroom or to make adequate use of the
resources in the room; b.) since necessary services such as restrooms
and nursing stations are generally distantly placed from most lounges,
residents may prefer to stay in their rooms or meet in the halls; c.)
the lounge provides no more interesting or entertaining stimuli than

some other areas such as the recreation room, hallway, or bedroom; 3) various forms of patient segregation, whether externally imposed by nursing unit regulations or by activity scheduling, or individually imposed due to greater freedom of the more able residents to move, may severely limit the physical and social behavior of many of the elderly; 4) interior designers and decorators must be attuned to the particular physical characteristics and needs of aging nursing home residents so that their efforts reinforce the overall goals of the facility; 5) specific items of furniture may have a profound effect on social behavior, tables prove to be useful in enduring all phases of social behavior: congregation, impersonal interaction, and conversation; and 6) if conversation is one means of reinforcing physical and social rehabilitation of the elderly, then guidelines for promoting meaningful social encounters between the elderly and between the elderly and staff members might be most helpful to nursing home administrators.

ULRICH, R.S.   244

"Visual Landscape Preference: A Model and Application," *Man-Environment Systems,* Vol. 7, No. 5, September, 1977, pp. 279-292.

KEY WORDS:

Man	Setting	Relational Concepts	Outcomes
Students, suburanities	*Sweden, U.S.*	*Cross-cultural, preference model*	*Similarities, dimensions*

The two-fold purpose of this study was to develop and test a model that relates aesthetic preference to informational properties of various photographs of landscape scenes. In following with the functionalist theories of S. Kaplan, the model was premised on the notion that evolution has left humans with perceptual and informational biases that affect visual preferences. Humans were viewed as creatures who need and like to acquire visual landscape information. The model hypothesized high preference for scenes having attributes that facilitate perception and comprehension, or which convey to the observer an anticipation that additional information can be acquired by altering the vantage point (movement through space). Various properties of perception were conceptualized as subjects' dimensions, which underlie their preferences. These dimensions were represented in the model as several component variables: complexity, focality, ground texture, depth, and mystery.

The model was tested by applying it to visual preference data obtained from two diverse groups of subjects: 1) students at a Swedish university, and 2) an older group of American suburbanites. Both groups were shown a collection of 53 photographs of rural roadside scenes in Michigan. An initial group of approximately 450 pictures were edited down to the number presented to subjects. Photographs were systematically taken to avoid possible bias caused by ephemeral features. Each subject individually rated each of the 53 photographs on a 6-point scale.

Results indicated that the model predicted both groups' preferences for the scenes with a high level of accuracy and communuality. Also,

results from factor analysis suggested that both the Swedish and
American subjects responded to properties in the scenes that were components of the model. Together the findings supported the validity of
the model, and strongly suggested that the theoretical framework cap-
tured some major determinants of visual preference.

The author stated that although the model held promise for applications to other environmental settings, i.e. urban and architectural, the
framework would become much more complex if the effects of water, color,
and living elements such as humans were accounted for in the model. The
need for more work in such applications was stressed. It was felt that
much of the previous work on aesthetic preferences for landscapes had
overstated the role of culture. While evidence exists in support of the
role of culture, recent surveys of cross-cultural research have
concluded that there is no indication that fundamental perceptual and
cognitive processes vary between cultures. In this respect, similari-
ties have been more prevalent than differences. It was concluded that
the role of informational determinants in aesthetic landscape assessment
must not be neglected.

VALINS, S. AND A. BAUM   245

"Residential Group Size, Social Interaction, and Crowding," *Environment
and Behavior,* Vol. 5, No. 4, December, 1973, pp. 421-439.

KEY WORDS:

		Relational	
Man	Setting	Concepts	Outcomes
Students	*Laboratory*	*Group size*	*Social interaction*

     The authors of this study hypothesize that students subjected to a
higher number of people in their living environment will result in a
higher degree of perceived crowding and an avoidance of social relations. Specifically, they propose that students living in dormitories
that share large, communal baths and lounges will express more social
isolation and stress than those living in suite arrangements which serve
only a small number of bath and lounge users.

     Freshmen at the Stony Brook campus of the State University of New
York were chosen as the experimental population of the study. Approximately 1,000 lived in corridor dorms in which 34 students shared common
bath and lounge facilities, while about 500 resided in suite dorms in
which only 4 to 6 students were required to use communal bath and recreation area. Groups of both sexes were randomly selected from this population in four groups of 64 each (32 from corridor dorms and 32 from
suite dorms). Each group was exposed to one of four laboratory experiments which were designed to measure stress and social interaction,
crowding, cooperation and competition.

     In support of the original hypothesis, it was found that a much
higher number of the corridor residents considered their living environment crowded (67% versus 25% for the suite residents). It was also
found that corridor residents met too many unknown people on their floor
and experienced a higher degree of social stress. Corridor residents

were more aware of negative features of the environment and were more prone to avoid social interaction. They would establish a greater social distance than suite dwellers and tended to talk to strangers less in social situations. Cooridor residents were also found to be more competitive than suite residents and tended to prefer situations which stressed solitary behavior. In conclusion, the authors found a significant degree of support to their contention that corridor residents felt crowded, developed isolated social tendencies, were exposed to too many unwanted social interactions, and experienced excessive degrees of stimulation.

VANSELOW, G.W.   246

"On the Nature of Spatial Images," *Proceedings of the Association of American Geography,* Vol. 6,   1974, pp. 11-13.

KEY WORDS:

Man	Setting	Relational Concepts	Outcomes
Random subjects	*University district*	*Horizontal and vertical components*	*Spatial changes*

Images, an individual's internal representation of the real world, include spatial components as well as temporal, operational, emotional and evaluative components. The formation and development of spatial images depends largely on an individual's experience with his environment and whether that experience is direct or indirect. The results of several past studies have indicated that subjects created maps which were either 1.) sequential, areas of the city connected by paths or roads, or 2.) spatial, a scattering of elements including buildings, landmarks, or districts. By analysis it is possible to relate the type of maps produced to the nature of the image, and ultimately, to the nature of the environmental experience.

A study was carried out to test the hypothesis that a significant proportion of individuals possess horizontal images of spatial environments. One hundred and fourteen subjects were selected at random from 21 census tracts surrounding University Way and the University district in Seattle. Subjects were presented with three maps and three sketches (ranging from vertical to horizontal) of a set of places of various scales. The five places selected were the United States, the State of Washington, the City of Seattle, the University District, and University Way. Subjects were asked which representation was most like the image they had for each place.

Analysis of the results indicated that a significant proportion of the population possesses horizontal images of spatial environments. Given the possibility that subjects saw all sketches as horizontal and all maps as vertical, responses were 296 sketch (horizontal) rsponses and 273 maps (vertical) responses.

Implications for imagery research suggest the redefinition of the terms mental map and spatial image. In addition, there is a need to develop techniques to compliment the use of sketch maps to examine the nature of spatial images. Finally, the planning and design professionals must consider the creation of more satisfactory future enviroments which require a greater concentration on elements such as local landmarks and skylines which are visible from a horizontal as well as vertical vantage point.

VERDERBER, S. AND G.T. MOORE   247

"Building Imagery: A Comparative Study of Environmental Cognition," *Man-Environment Systems,* Vol. 7, No. 6, November, 1977, pp. 332-341.

KEY WORDS:

Man	Setting	Relational Concepts	Outcomes
Urban dwellers, students, suburbanites	*Urban area*	*Comparative building imagery, cognition*	*Dimensions, preferences*

Meaning in architecture was conceptualized as the systematic communication of intrinsic messages or signs that convey fragments of information from object to perceiver. Various types of meanings are influenced by both the social-economic context and background of the individual or group in question. Laypersons' descriptive appraisals of the built environment have often been either intuitively assumed by the architect or acknowledged in an informal manner. This study sought to contribute towards minimizing the information "gap" which was felt to exist between the values, needs, and desires of three groups of non-architects as opposed to theoretical intentions of architects. Building imagery was viewed as the study of relationships between persons' enviromental cognition and how resultant dimensions of imagery influence architectural preferences.

Objectives of the study were to comparatively measure: 1) cognitive building images across socio-economic and lifestyle groups; 2) the aesthetic impact of a set of specific buldings and the subsequent preferences as seen by subjects between groups, 3) techniques for eliciting hard (formal) data as opposed to soft (informal) data, and 4) the potential applicability of results to the architectural design processes.

An aesthetic dichotomy was developed that differentiated High from Popular Architecture. This theoretical framework for the classification of objects in the built environment was holistic in nature. This system allowed for the discussion of a wide range of both historical and contemporary buildings. In this study, Architecture with a capital "A" (buildings that possess serious commodations) was considered equally significant to architecture with a small "a" (buildings that are ubiquitous or somehow insignificant). The assumption was made that both types of buildings must be responsive to behavioral needs, as well as the aesthetic aspirations of persons. Five hypotheses were tested that focused on behavioral dimensions of building imagery.

A comparative study of three distinct respondent groups located within the Milwaukee metro area was conducted. Respondents represented a urban lower-income area (N = 21), a suburban upper-income area (N = 21), and an undergraduate group of non-architecture university students (N = 21). Each respondent was presented with ten sets of color photographs depicting five buildings pre-classified as Popular Architecture and five buildings pre-classified as High Architecture. Each building was represented through five views ranging from overall (1) to detailed in scale (5). Buildings were unfamiliar to respondents. A pilot study (N = 15) was conducted to pre-test measures prior to the formal study.

Response measures for each of the ten buildings consisted of: 1) a building imagery semantic differential, 2) six-point architectural aesthetic preference scales, 3) free written building descriptions, and 4) respondent background data. These items were presented in a questionnaire. The principal objective of the pilot study was to edit an initial list of 63 pre-determined bi-polar scales to only those scales deemed most salient by respondents relative to the ten buildings. In the formal study the four response measures were presented. Counterbalancing was incorporated to minimize response bias within groups.

Objectives of data analysis stages were to identify: 1) dominant empirical dimensions of building imagery, 2) architectural aesthetic preferences between buildings and between groups, 3) links between the non-theoretical continuums and subsequent empirical dimensions of building imagery, and 4) contents of the free written building descriptions and potentially augmenting the dimensions of imagery.

Major results of these measures were: 1) the original list of 64 bi-polar semantic males was reduced to 41 which were regarded as salient across groups, 2) factor analysis results yielded eight principal dimensions of imagery across groups and buildings, 3) each of the three groups were more homogeneous internally than they were in comparison with other groups, 4) clear-cut differences in factor analyzed imagery dimensions were noted between groups, and to a lesser extent between buildings, 5) preference results indicated that High Architecture buildings were preferred over the five Popular Architecture buildings across groups, and 6) the written building description analysis yielded sets of complimentary, subjective images that focused on physical attributes and levels of detail. It was concluded that group differences were clearly discernable between groups and buildings. Additionally, architectural meanings and imagery are both quantitatively and qualitatively measurable. The authors stressed that further research should focus on the development of additional continuous sub-dimensions of imagery, and applications to other respondent groups.

WALLER, R.A.   248

"Environmental Quality: Its Measurement and Controls," *Regional Studies,* Vol. 4, 1975, pp. 177-191.

KEY WORDS:

Man	Setting	Relational Concepts	Outcomes
Human	*Living environments*	*Measurement and annoyance*	*Environmental quality*

Given the present state of technology almost any level of quality within the physical environment can be achieved. In the past, however, the balance between improvements in environmental quality and their costs have not been studied in great detail. It is felt that if we are to control our environment we must find a measure of the physical environment that correlates with peoples' enjoyment of it. We must then evaluate the work of the environment in monetary terms in order to judge how much to spend. This paper proposed techniques of evaluation and in addition proposed a model whereby the many diverse aspects of our environment can be related to a common monetary scale.

Many references are cited on the measurement of the main aspects of the external physical environment. The decision model for enviromental control is described in a five step process: 1) describe the environment in physical terms, 2) forecast effect on people, 3) evaluate (i.e. how much is it worth paying to reduce complaints?), 4) compare (i.e. how much does it cost and how much is it worth?), and 5) decide (i.e. which solution is acceptable in terms of all costs?).

The common scale set forth to which all measured aspects of the environment can be related is an annoyance scale. From this combined annoyance scale one can calculate the "loss of amenity" in monetary values/household by the use of regression techniques and "normal probability" scaling. Examples of aspects of the environment which are shown as related to annoyance and "loss of amenity" are; motorway noise, airport noise, lamp flicker, door height, and party walls. It is proposed that such a model would fit all aspects of the external physical environment. The use of such an approach can have significant impact on public policy, planning and design decisions.

WARNER, P., J. JACKSON, L. GOOD, AND A. GREENBURG   249

"Measurement of the Influence of a Cement Kiln Stack on a Surrounding Residential Community by Injection of an Indentification Particulate," *American Industrial Hygiene Association Journal*, January, 1975, pp. 32-38.

KEY WORDS:

Man	Setting	Relational Concepts	Outcomes
Man	Residential	Effluents/ pollution	Effluents effect

In this study, an identification particulate, barium sulfate, was injected into the kiln stack of a large cement plant in order to determine the effect of the particulate effluent of the stack on the immediate surrounding residential community.

Settleable particulate was determined by evaporation of the contents of dustfall samplers with subsequent analysis performed by atomic absorption spectrophotometry.

Suspended particulate was collected from high-volume sampler stations. Tared filers were desicated to constant weight, and total particulate loading was determined by using an air volume specific to each sampler as ascertained by rotometer standarization.

Results of dustfall exposure indicated little evidence of detectable barium tag-particulate over the 24 hours following the stack injection. A slight measurable effect of the kiln stack emissions at the Detroit-based cement company is observed at ground level in suspended particulate at a distance of 1.6 km from the stack. Stack related particulate was observed at no other sample site within the environmental area chosen for study.

WEINSTEIN, N.D.   250

"Individual Differences in Reactions to Noise: A Longitudinal Study in a College Dormitory," *Journal of Applied Psychology*, Vol. 63, No. 4, August, 1978. pp.458-466.

KEY WORDS:

Man	Setting	Relational Concepts	Outcomes
University students	Dormitory	Noise/longitudinal analysis	Sensitivity judgements

This two part study investigated differences among individuals in their initial reactions to noise and in their ability to adapt to noise over a period of time. The general aim was to assess students' sensitivity to noise before they arrived on campus and then to compare their initial reactions to the dormitory noise with the final level of adjustment reached at the end of the school year. Also, personality correlates were obtained from five samples in Study Two in order to investigate possible origins of sensitivity to noise.

In Study One, college freshmen (N = 155) completed a self-report measure of noise sensitivity (prior to the school year) and two subgroups were constructed from students whose noise-sensitivity scores fell within either the top or bottom 30% of this group. Each group had nearly equal numbers of males and females. Self-reports of dormitory noise disturbance were obtained from the noise-sensitive group (N = 24) and noise-insensitive group (N = 31) early in the school year and again seven months later. The rooms of the students were located throughout the six wings of a coed dormitory at Rutgers University. The building was of a typical, double-loaded corridor design with double occupancy rooms.

The research instrument consisted of a questionnaire that elicited a wide array of information pertaining to the assessment of self-reported sensitivity to noise and the measure of noise disturbance. Typical questions inquired as to how often the student asked a neighbor to make less noise and how annoying he or she found the dormitory noise when trying to sleep. Other issues were analyzed, such as privacy, roommate relations, storage space, and having to share washrooms. A principal-components factor analysis was used to analyze data. In Study Two, data was gathered via a series of personality tests.

Results indicated that noise-sensitive students were much more bothered by dormitory noise ($p < .0001$) and that they became increasingly disturbed during the year ($p < .01$). Noise-insensitive students showed no change. Correlations of the noise-sensitive scale with academic test data and personality inventories suggested that noise-sensitive students are lower in scholastic ability, feel less secure in social interactions, and have a greater desire for privacy than their less noise-sensitive peers. This research was viewed as having practical implications for student housing. Not only was noise reported to be the most unpleasant feature of dormitory living, but certain groups of students apparently are hampered by noisy environments. Changes in dormitory design and policy raises the possibility of reducing stress and interpersonal conflict by placing students together whose reactions to noise are similar.

WHEELER, J.O. AND F.P. STUTZ   251

"Spatial Dimensions of Urban Social Travel," *Annals of the Association of American Geographers,* Vol. 61, No. 2, June, 1971, pp. 371-386.

KEY WORDS:

Man	Setting	Relational Concepts	Outcomes
Residents	*Urban area*	*Distance decay, social trips, socioeconomic status*	*Social trip patterns*

One major function of an urban area is to promote and maintain linkages among its different activity areas. These linkages include: 1) activity-to-activity ties, 2) person-to-activity connections, and 3) person-to-person interactions. Although previous studies have examined the first two types of linkages, a review of the literature revealed virtually no analyses of social trips. Consequently, two hypotheses were tested by the authors: 1) the frequency of social trips will decline with increasing travel distance between the interacting participants, and 2) an individual of a given socioeconomic status will be more likely to choose as a social contact someone of similar status, his frequency declining with progressively dissimilar status levels.

Basic data was obtained from a 1965 home interview survey of over 4,500 households in the three county Lansing, Michigan metropolitan area. Over 63,000 social trips (those trips having a residential destination) were analyzed across a spectrum of demographic variables.

An examination of the survey data revealed that the distribution of social trips varied by age, occupation, sex, and race. Compared to total trips, social trips made up a disporportionately large share of trips among the twenty to forty-five age group, and for those in their early years of retirement. After the mid-forties social trips were a decreasing purpose for travel, and after seventy they tended to decrease rapidly with age. Professionals were found to have a smaller proportion of social trips to all trips than those occupational groups at the lower end of the socioeconomic scale. In addition, women were found to account for 54% of all social trips. Finally, blacks were more highly represented in social trips to all trips than were whites.

The findings also provided support for the hypotheses that distance between participants was meaningful, and that distance separation was related to socioeconomic status. Thus, individuals tend to be constrained in terms of social travel by perceived social barriers based on status levels, and limited in a spatial sense due to distance or travel time. These findings led the authors to suggest that "it is likely that neighborhoods in social equilibrium would have a high degree of local social intimacy and neighborhoods of diverse makeup would have fewer local ties." They further suggested that, "although not studied here, hypotheses that might be examined in the future are the similarity of ties among like age groups and among families of similar demographic characteristics.

WHITE, S.E.   252

"Residential Preference and Urban In-migration," *Proceedings of the Association of American Geographers,* Vol. 6, 1974, pp. 47-49.

KEY WORDS:

Man	Setting	Relational Concepts	Outcomes
Urban dwellers	*Cities*	*Residential preferences*	*Urban in-migration*

This investigation sampled residents at the state-wide scale in order to ascertain relationships between residential preference and urban in-migration. Most past studies have relied on student populations and attempted to assess preference at a national level. This approach attempts to overcome such potentially bias causing techniques by sampling at a broader scale while questioning subjects about cities with which they were likely to have more first-hand knowledge.

Four hundred thirty-five individuals responded to mail questionnaires. Age and income characteristics were representative of the state-wide situation. Respondents generally had a stronger educational background. Respondents were asked to rate 25 cities on a five point scale from desirable to undesirable as a place of residence. They could indicate a lack of knowledge from which to judge. In addition to certain demographic data (age, income, education, etc.) they were asked which cities they had visited and to rank in order of importance factors they considered most prominent in influencing in-migration.

Maps developed from the preference data and urban in-migration data (calculated from the 1970 census) were highly comparable. The Pearson correlation coefficient between residential preference and urban in-migration was +.67. Almost half of the variation of in-migration can be explained by the variation in residential preference attitudes. When combined with other commonly accepted migration attraction indicators, such as personal income per capital, industrial employment per capital, city size, knowledge, visitation, and median family income in a simple correlation table and stepwise multiple correlation 74.2% of the total variance was explained. All but two of the variables, city size and industrial employment, were significantly related to urban in-migration ($p < .05$). Visitation and knowledge were strongly associated to urban in-migration suggesting the importance of the "mental image."

Preference factors gathered from respondents were tabulated by age. Results indicated that residential preference criteria changed between different groups of individuals. While the young placed great significance on more pay (46.4% ranked it first), middle age people rated climate and more pay almost equally (31.2% and 28.8%), and the elderly were primarily interested in climate (46.4%). Nearness to relatives became increasingly important for the elderly (18.3%). Other factors were ranked fairly consistently across groups.

In conclusion, the author felt that there was evidence to support the development of a preference based inter-urban in-migration model.

WHITEHEAD, B.   253

"The Rational Planning of School Layouts," *Build International,* Vol. 3, No. 12, December, 1970, pp. 377-380.

KEY WORDS:

Man	Setting	Relational Concepts	Outcomes
School staff, students	Secondary school	Design determinants, routine journeys	Design recommendations

Despite a 1954 edict by England's Minister of Education that movement in schools should be analyzed carefully so that it can be taken into account during the planning of school layouts, no comprehensive survey of movement in a school building was known to have been published. Consequently, the author utilized a computer program to examine the relationship between numbers and types of journeys, and building design. In addition, factors other than human movement were also integrated into the planning process in order to encompass a wider spectrum of considerations.

A relatively new Secondary Modern School was chosen as the site for the study. A new building was chosen for two major reasons. First, the fact that the school had been recently designed meant that changes in its initial function were not likely to have occurred yet. Second, it would be easier to access and appraise the decisions which had been made in planning the building.

Collection of the data took two forms: 1) extracting what movement had already been recorded in the school timetable, and 2) a sampling survey covering 40 working days. Numbers of journeys between 40 points were recorded for each of eight categories of people. Weighting of the "importance" of an individual member of each group was done based on the salaries of staff and the costs of education of pupils. Results were then presented in terms of the numbers of journeys at each representative period of time, between each pair of activities, and for each category of people.

Movement of the janitors, visitors, inspectors, kitchen staff, and the secretary was comparatively small. When standard movement is taken into account the dominant patterns are shared with pupils and teachers.

The results of the computer analysis were summarized as follows: 1) the number and strength of links was particularly great in the administrative and communal spaces, e.g. head teacher's office and dining hall, 2) there were also strong links between external entrances, dining hall, assembly hall, and playground; playgrounds and toilets; head teacher's office, staff room, and staff toilet and 3) the amount of movement between classrooms was small. When the movement data was combined with nuisance restrictions, the following recommended features were suggested by the analysis: 1) the noisy practical rooms should be grouped together and cut off from the quiet areas by the art room, toilets, and open spaces, 2) the playground should be located next to the assembly hall away from the quiet administrative and classroom areas, and 3) the classrooms need to be clustered together more than when planning for movement alone.

WINETT, RICHARD A.   254

"Prompting Turning-Out Lights in Unoccupied Rooms," *Journal of Environment Systems,* Vol. 7, No. 3, 1977-79, pp. 237-241.

KEY WORDS:

Man	Setting	Relational Concepts	Outcomes
Students	*Classrooms*	*Reinforcement signage*	*Unambiguous cues*

The author of this study proposed that reinforcement signage in the environment must be strategically located and specific in its message and intent. He argued that comparisons between signage systems with only subtle variations are discernible and the effectiveness of the more specific cues can be predicted.

Three classrooms were used as the setting in this study, two being used as control units and one as the experimental model. Lights in each room were controlled by panels in readily accessible locations and convenient to the use of occupants leaving. Observations were made each day over a six week period and the number of times the lights were left on in each setting were recorded. A baseline of a two week period measured the percentages of days the lights were found on with no signage in any rooms. At the three week point, the author introduced the university's standard energy conservation sign and during the fourth week added a university conservation sticker in the experimental setting. During the fifth and sixth weeks, he removed the standard signs and installed a number of specific cues which requested that the lights be turned off in the experimental room at a specific hour.

The lights in the control rooms were left on for 78 percent of the time during the test, while the lights remained on in the experimental unit for 95 percent of the time, even with the introduction of the general university conservation signs. Only during the fifth and sixth weeks did the experimental room show a decrease in the amount of time that the lights were left on (40 percent). The author concluded that the only effective behavioral reinforcement came with the institution of a system that conformed to his hypothesis that people will respond only to specific and unambiguous cues.

"Space Colonization--Some Physiological Perspectives," *Aviation Space and Environmental Medicine,* Vol. 49, No. 7, July, 1978, pp. 898-901.

KEY WORDS:

Man	Setting	Relational Concepts	Outcomes
Space colonizers	Space colonies	Adaptation/space environment	Physiological constraints

Although the establishment of large self-sustaining communities in free space is possible with present day technology, our understanding of the physiological design criteria determining habitat mass, structural configuration, cost, and ultimate feasibility, is inadequate.

This study summarizes the biomedical findings of an engineering workshop cosponsored by the American Society of Engineering Education, Standford University, and NASA's Ames Research Center in 1975, with special emphasis on recommendations for further research into relevant environmental physiology problem areas.

Although the human body maintains a remarkable capacity for adaptation to terrestrial variations and stresses, living in free space represents a significant challenge to some of these homeostatic mechanisms. The internal milieu requires protection from injurious physical agents (e.g. radiation, micrometeorites) present in, and provision of vital factors (e.g. atmosphere, gravity, food, water, oxygen, diurnal periodicity) absent from the external environment.

This paper emphasized that such protection and provision to date has been limited to highly selected astronauts working in space for comparatively brief periods of time. We have virtually no experience in engineering a space habitat for a large, permanent, heterogeneous population. The author suggested that where our knowledge of potentially dangerous effects is incomplete, a conservative design policy must be applied.

It was also suggested that before large human communities in free space can be established, research is necessary to determine optimal levels in three critical areas, gravity, atmospheric composition and pressure and radiation protection.

The author concluded that a viable, economically productive space community of 10,000 people could be established with earth-normal gravity, general population maximum radiation exposure standards, and a reduced nitrogen partial pressure.

"Survival in Relocation," *Journal of Gerontology,* Vol. 24, No. 4, July, 1974, pp. 440-443.

KEY WORDS:

Man	Setting	Relational Concepts	Outcomes
Elderly	*Housing*	*Relocation, mortality rates*	*Positive influence*

Previous studies of the increased mortality rates of institutionalized patients following relocation have led some to conclude that it might be best to administer treatment of the elderly in their own homes rather than to have them move to institutions. There has been evidence that involuntary relocation is detrimental to longevity when the elderly person is suffering from an illness prior to the move. When the elderly person has been healthy, a voluntary move has not always been viewed as endangering the person's chances for survival. In a prior study Lawton and Yaffe found no significant difference in mortality rate between a group of relatively independent elderly persons who voluntarily moved to a congregate housing development and a group of matched nonrelocated residents.

In the present study, mortality rates were compared between older healthy community residents who relocated voluntarily to new apartment complexes and control group residents who did not relocate. Also, actuarial-type data were compared between relocated subjects who lived and those who died within one year following the move.

The subject group consisted of 732 applicants to two federal elderly housing developments. Average age of subjects was 74. Buildings 1 and 2 were similar. Applicants had to be sufficiently healthy. Movers were compared to non-movers in terms of death rate and movers were compared to the elderly population at large. Demographic comparisons between groups were made. A series of 15-point health rating scales were completed for each applicant in consultation with a physician. Most subjects possessed scores ranging between 12 and 15.

Two hundred forty-three people moved into Building 1 and 219 persons moved into Building 2. Evidence did not suggest that the lives of these elderly persons were shortened by the stress of relocation. The percent of the movers who died within one year in each of the buildings 1 and 2 was 3.7. Contrasting to this low death rate, 5.4 percent of non-mover applicants to Building 1 died in the same period. An analysis of six demographic characteristics for each group yielded one significant difference between death rates and marital status.

It was concluded that although the risk of relocation appeared to have no increased effect on the death rate for these elderly persons, previous studies have shown that an involuntary move to a nursing home has often caused sick and highly stressed persons severe hardship that ultimately resulted in death. It has been concluded by others that such persons respond poorly to forced relocation. Sex has been described as an additional factor contributing to increased mortality rates following relocation (males had died sooner than females). The proper planning and design of residential facilities for the elderly possesses the power of allieviating many negative effects on elderly persons, particularly in cases of relocation to new facilities.

WOHLWILL, J.F., J.L. NASAR, D.M. DEDOY AND H.H. FORUZANI   257

"Behavioral Effects of a Noisy Environment: Task Involvement Versus Passive Exposure," *Journal of Applied Psychology,* Vol. 61, No. 1, Febuary, 1976, pp. 67-74.

KEY WORDS:

Man	Setting	Relational Concepts	Outcomes
University students	*Room*	*Noise exposure*	*Activity patterns/ tolerance*

The voluminous body of experimental research concerning the effects of noise on behavior suggests that individuals may be able to tolerate fairly high intensity noise without appreciable impairment of their performance on diverse kinds of tasks. This study compared the effects of noise under active task involvement as opposed to passive exposure. The two major hypotheses tested were: 1) imposing several overlapping sources of continuous auditory stimulation on subjects engaged in a complex task will not affect task performance but will be manifested in lowered resistance to frustration following such exposure, and 2) such auditory stimulation will not result in lowered resistance to frustration on the part of subjects only passively exposed (to stimulation) and they will not experience stress induced by more difficulty in task performance.

Eighty undergraduate subjects were assigned to one of four conditions; representing 2 x 2 combinations of task versus no task, and noise versus quiet. The experiment consisted of two parts. The auditory stimulation for a dial-coding consisted of recorded music of varying types, a potpourri of office sounds, and taped interviews. A 30 x 30 foot room was the setting for the experiment. A set of mood scales were administered twice to all subjects. They were asked to describe present feelings on each of 12 rating scales. Slides were used to augment the auditory stimulation and to provide a basis for the rating scales. For part two of the experiment, each subject was asked to solve four puzzles based on line-tracing pictorial patterns.

Performance on the dial-monitoring task was unaffected by noise. Ratings of interest and tenseness were significantly higher under the task condition (tenseness was also higher under noise). On a test of resistance to frustration both noise groups (regardless of task condition) showed a smaller degree of persistence on insoluble puzzles than the no-noise groups. These results suggest that after-effects of noise are not dependent on the power of noise to disrupt task performance. Thus, the first hypothesis was confirmed while results based on the second hypothesis do not entirely confirm its validity.

The authors conclude that this discrepancy suggests that the effect of prior auditory stimuli on subsequent problem solving performance may operate well below the subject's level of awareness, and in a fashion quite independent both of the subjects prior adaptation to the stimulation and of subjects expressed affective state. It was viewed that this conclusion is important in many ways, given the pervasiveness of extraneous auditory background stimulation in everyday activities.

WOLFE, M.   258

"Room Size, Group Size, and Density: Behavior Patterns in a Chldren's
Psychiatric Facility," *Environment and Behavior,* Vol. 7, No. 2, June,
1975, pp. 199-224.

KEY WORDS:

Man	Setting	Relational Concepts	Outcomes
Children	Psychiatric hospital	Room size, group size, density	Density factors

Observational data were collected in a children's psychiatric facility over a period of 2 1/2 years. The objective was to further understand the use of bedrooms and the relationship between room size, group size, density and associated patterns of use and behavior. Relevant psychiatric facilities design literature was discussed. The hospital facility and subject group was described in detail.

In order to study the use of space in the hospital, a series of four observational studies were conducted. Data for the present paper were obtained by combining bedroom-use mapping techniques. An observer made a predetermined tour every 15 minutes, recording ongoing activity, residence status, sex of patients, and the number of staff present. Bedroom use was evaluated by three criteria: 1) percentage of occupied periods, 2) average number of children present when rooms were occupied, and 3) percentage of room use per child. Combining all rooms over the entire observation time span, there were 5,951 periods. During the 776 occupied observations, a total of 854 children were observed in 781 activities.

Various combinations of room size (49.52, 118.05, and 220.99 sq. ft.) and group size (1, 2, 3, or 4 children) occurred naturally, yielding densities ranging from 29.01 sq. ft. per child to 220.99 sq. ft. per child. Findings indicated that the definition of potential density (sq. ft. per person if all were present) which is widely used in the majority of architectural programming endeavors, was not related either to bedroom use or behavioral patterns. Rather, room size and group size were significant, not as parts of a mathematical density factor, but as they interacted to create specific psychological density conditions. Specific design and administrative implications of the findings were presented.

"Study of Office Equipment - Attitudes to Office Landscapes and Open-Plan Offices," *Build International,* Vol. 6, No. 1, January-February, 1973, pp. 143-146.

KEY WORDS:

Man	Setting	Relational Concepts	Outcomes
Employees	*Office buildings*	*Attitudes*	*Work setting preferences*

Efforts have recently been made to utilize modern methods of office landscaping and open-plan design, based on the assumption that these environments will promote a more pleasant work experience. A study of approximately 2600 persons from 38 different companies in Sweden employed a questionnaire designed to compare the attitudes of employees in four "modern" office types to those employees working in conventional offices.

Attitudes were compared between the following four types of offices: 1) large office landscapes, new offices designed and built as office landscapes, each having an effective office area of a least 800m^2; 2) smaller office landscapes, areas 250-800m^2, otherwise the same description as above; 3) large open-plan offices; converted buildings, the office area at least 800m^2; and 4) smaller open-plan offices; converted buildings, the office area between 250-800m^2.

Results revealed significant attitudinal differences between employees of the four types of offices. In general, there were more positive attitudes towards smaller office landscapes, followed by larger landscapes, and then the two other office types. Although acoustics and ventilation were criticized regardless of office design, more effective acoustic situations were found in the smaller landscapes, and more effective ventilation found in the larger landscapes. In terms of social and psychological effects, 35-55 percent felt more tired, 30-40 percent more tense and irritable, and 35-50 percent more disturbed than before moving from conventional offices. Collaboration and communication was found to increase after moving from the conventional offices in the majority of cases. The author concluded that, "if the office landscape system is to be adopted by an organization, a wholehearted effort must be made to create the best possible environmental conditions."

"Psychological and Social Correlates of Life Satisfaction as a Function of Residential Constraint," *Journal of Gerontology,* Vol. 31, No. 1, January, 1976, pp. 89-98.

KEY WORDS:

Man	Setting	Relational Concepts	Outcomes
Elderly	*Residential*	*Constraint, satisfaction*	*Low constraint preferences*

Previous research has examined various correlates of elderly persons' life satisfaction levels. Life satisfaction has been viewed as an important component of successful aging. These correlates have usually included "health, activity level, formal and informal group participation, education, and income levels." This study acknowledged the value of addressing these several social-psychological variables coupled with inquiry as to whether various environmental settings contributed to varying levels of life satisfaction. Two questions were addressed: 1) did residential settings of varying levels of constraint influence life satisfaction, and 2) did health, developmental task accomplishment, self-concept, activity level, and perceived autonomy differentially predict life satisfaction in such settings?

A total of 129 ambulatory patients were surveyed in settings of high and low environmental constraint on the following measures: 1) life satisfaction, 2) task resolution, 3) self-acceptance, 4) autonomy, 5) activity level, 6) health, and 7) educational level. Both residential settings (A and B) were located in suburban areas. Setting A residents had private rooms although basic needs were met through staff functions. This environment posed considerable constraints on residents. Setting B residents lived in a "retirement-village" where privacy was important and all basic needs were met individually. Fewer constraints on behaviors were imposed here. Detailed profiles of both settings were provided by the authors. The series of research instruments were administered to each subject in the form of rating scales. Each portion of the overall questionnaire coincided with one of the seven specific variables being addressed.

Stepwise regression, covariance, and t-test analyses indicated: 1) life satisfaction and developmental task accomplishment were greater in the lower constraining setting, 2) the selected correlates resulted in multiple correlations of .675 and .590 with satisfaction, 3) differing sets of correlates significantly predicted satisfaction in each setting. Health was most important in the high constraint setting and perceived autonomy and self concept was important in the low constraint setting, and 4) developmental task success significantly predicted satisfaction in both settings.

The less constraining residential environment was perceived as more likely to "foster or maintain personal growth and development." This coincides with the authors' view that feelings of many persons in institutionalized settings are such that high constraint environments strongly contribute to this acute loss of independence. All of the variables measured were discussed in relation to both high and low constraint residential settings.

The authors concluded that: 1) previous research which has considered satisfaction and adjustment during the aging process has failed to define specifically the nature of the setting from which samples were drawn and to assess the influence of setting differences upon satisfaction and its determinants and 2) other situational parameters related to social and psychological processes need to be considered in further investigations.

"The Effect of the Meaning of Buildings on Behavior," _Applied Ergono-
mics_, Vol. 1, No. 3, June, 1970, pp. 144-150.

KEY WORDS:

Man	Setting	Relational Concepts	Outcomes
Housewives	_Scotland_	_Meaning, behavior_	_Perception_

The authors of this article proposed that subtle differences in a
person's perception of an environment can be measured and cataloged
using standard semantic scales. The purpose of this study is the experimentation of this thesis using established techniques of psychological
measurement.

Thirty-two Scottish housewives aged 21 to 35 were used as subjects
for this experiment. A set of 24 monochromatic line drawings and a pair
of black and white photographs which were pretested for accuracy were
used as the test environment to be evaluated. A series of semantic
scales had previously been established which were designed to measure
subjects' attitudes within such areas as "friendliness", "harmony", and
"activity". The subjects were shown the sets of pictures and asked to
respond to them using the scales provided by the researchers. The
research design focused on attitudes about the environments that could
be obtained from the subjects in such a way that generalizations about
the appropriateness of each setting could be established.

Using varimax rotation analyses techniques on data collected, it
was shown by the authors that high correlations in the subjects'
responses could be drawn to the various types of environments presented.
In both studies, as well as in the pretest experiments involving student respondents, the team was able to establish a consistent match
between specific settings and subject response. Measures of "appropriateness" were established for each environmental setting based upon the
results of the experiment and generalizations about each environment
were outlined. The authors concluded that, using the semantic scales
established in this and previous studies, environments can be evaluated
in a consistent and statistically meaningful fashion for real building
settings as well as the pictorial representations presented herein.

"Personal Space as a Function of the Stigma Effect," _Environment and
Behavior_, Vol. 6, No. 3, September, 1974, pp. 289-294.

KEY WORDS:

Man	Setting	Relational Concepts	Outcomes
Passersby	_Airport lobby_	_Personal space_	_Stigma effect_

The hypothesis of this study was that personal space would be increased by subjects when the experimenter had a highly visible stigma, or spoiled identity; and that subjects would increase their interpersonal distance on encounter with the experimenter as a result of a contamination effect. Personal space was defined as "the invisible boundary surrounding an individual into which another may not come." Personal space, sex differences, and time spent in encounter were measured in an airport lobby with randomly chosen subjects, simulating a stigma. Identical factors were recorded in a randomly selected control group by the non-stigmatized experimenter.

Twenty-nine control subjects were conducted by a "normal" experimenter, asking for freeway directions, and distances of approach were measured accordingly. The control group testing occurred one week after the initial test. The two tests occurred in similar conditions and in the same location. Thirty-four subjects were contacted by the stigmatized experimenter, wearing a tube in her nose and seated in a wheelchair. Distances of approach were measured during the encounter.

It was found that distances were significantly increased in interaction in the experimental situation over the control situation. The hypothesis was supported that the stigma effect operates to change interpersonal space preferences of individuals. The author viewed the endeavor as a heuristic study. Future studies were urged in other settings using the same research design to view personal space deviations among subjects under these two types of conditions.

YEE, W. AND M.D. VANARSDOL   263

"Residential Mobility, Age, and the Life Cycle," *Journal of Gerontology,* Vol. 32, No. 2, March, 1977, pp. 211-221.

KEY WORDS:

Man	Setting	Relational Concepts	Outcomes
Families	*Residences*	*Mobility, age, life cycle*	*Predictive mobility model*

This study presented a life cycle explanation for residential mobility. Relationships between events, age, and attitudes of family units and individual family members were discussed via the literature and these prior studies were evaluated accordingly. It was the authors' view that age-related events in a normative content influence moving probabilities for homogeneous populations who have relatively uniform social patterns. The life cycle was expressed mathematically by an a priori step function where transition points were delineated. The points were then used simultaneously with an assumed underlying distribution where a = age. Curve fitting was used to determine if the life cycle explanation was consistent at different units and levels of analysis. The following hypotheses were tested: 1) Persons' realistic possibilities and choices of where to move influence the individuals mobility plans prior to moving. Plans are thus linked more closely to subsequent residence changes than are choices. 2) If hypothesis one is true, then distinct family life cycle stages (steps) are apparent in age distributions of moving plans for homogeneous aggregates. 3) Family life cycle stages are less pronounced in the mobility patterns of heterogeneous than for homogeneous aggregates. The first hypothesis was tested for individual behavior. The second hypothesis tested for the population of an urban area (Los Angeles - Long Beach) and the third made a comparison between this metro area and a large sample across the U.S.

A series of age-related events were categorized ranging from 19 (beginning age) to 64 (death either spouse) which were ascertained via subjects' demographic profiles. A questionnaire contained items that differentiated between movers and stayers according to nine fractional component variables.

Results indicated that: 1) Age had an inverse relation to mobility. 2) Age was the underlying component of a family life cycle step function. This explained the mobility patterns of the homogeneous population aggregate. 3) Family life cycle transition points most likely delineated normative events (marriage, death, birth of children, etc.) that were followed by subsequent residential changes. The authors suggested that further research should focus on the transition points and how they might act as predictive time variables where behavioral probabilities are integral to the model. Also, a series of limitations and inconsistencies pertaining to the study were discussed. It was concluded that the indexing of family life cycle patterns in aggregate terms precludes the analysis of valuable data concerning a typical behavior which occur on an individual basis, and that the majority of "explanations of aging will continue to vary to the extent that they rest on theories assuming differing social clocks."

Key Word Indexes

AUTHOR YEAR	ABSTR. NO.	MAN	SETTING	RELATIONAL CONCEPTS	OUTCOMES
AIELLO, J.R. & T.D. AIELLO, 1974	1	Children	Schools	Proxemics	Social Behavior
AIELLO, J.R., D.T. DeRISI, Y.M. EPSTEIN & R.A. KARLIN, 1977	2	Female students	Laboratory	Crowding	Stress
ALLEN, M.A. & G.J. FISCHER, 1978	3	Male students	Laboratory	Thermal comfort	Learning performance
ALLEN, T.J. & P.G. GERSTBERGER, 1973	4	Office workers	Environmental arrangement	Territori- ality	Communication/ performance
AMICK, D.J. & F.J. KVIZ, 1975	5	Human	Public housing	Density	Social aliena- tion
ANDERSON, I., G.R. LUNDQUIST, & D.F. PROCTOR, 1973	6	Male students	Thermal environment	Thermal comfort	Humidity sensa- tions
ANDERSON, I., G.R. LUNDQUIST & L. MOLHAVE, 1975	7	Human	Indoors	Pollution/ building material	Contamination
ARCHEA, J., 1977	8	Human	Built envir- onment	Privacy	Environmental analysis
ARVIDSSON, O. & T. LINDVALL, 1978	9	Male students	Sonic environment	Noise	Task perform- ance
ASHTON, R., 1971	10	Infants	Hospital laboratory	Noise/illum- ination	Comparative levels response
ANUOLA, S., R. NYKYRI & H. RUSKO, 1978	11	Workers	Factory	Occupation	Strain
BAIRD, J., B. CASSIDY, & J. KURR, 1978	12	Students	Rooms	Architec- tural elements	Preferences
BAKEMAN, R. & R. HELMREICH, 1975	13	Research teams	Undersea habitat	Group cohe- siveness; performance	Performance in- influences cohesiveness
BALDASSARE, M., 1975	14	Urban dwellers	Densely popu- lated city area	Densely popu- lated environ- ment/social behavoir	Low neighboring
BALLANTYNE, E.R. & J.W. SPENCER, 1972	15	Tropical householders	Port Moresby, New Guinea	Thermal factors build- ing materials	Thermal comfort

AUTHOR YEAR	ABSTR. NO.	MAN	SETTING	RELATIONAL CONCEPTS	OUTCOMES
BALTES, M.M. & C. HAYWARD, 1976	16	Spectators	Football stadium	Littering control	Nonlittering behaviors
BANZIGER, G. & K. OWENS, 1978	17	Mental patients	Institution; two nonurban areas	Maladapta- tion; social indicators, geographical variables	Weather factor predictors
BATCHELOR, J.P. & G.R. GOETHALS, 1972	18	Students	Laboratory	Collective decisions	Spatial arrange- ments
BAUM, A. & S. VALINS, 1974	19	Students	Laboratory	Crowding/ residential environment	Values-issues problems/experi- mental data, use of
BAUM, A., M. RIESS, & J. O'HARA, 1974	20	Students	Drinking fountain	Personal space	Invasion
BECHTEL, R.B. & A. GONZALEZ, 1971	21	Psychiatric patients	Psychiatric hospitals, Peru, U.S.	Treatment settings	Behavior modifi- cation
BECKER, F.D., R. SOMMER, J. BEE, & B. OXLEY, 1973	22	College students	Classrooms	Environmental arrangement	Social behavior
BECKER, F.D., 1976	23	Apartment dwellers	Housing	Architectural arrangement	Children's behavior
BELCHER, J.C. & P.B. VAZQUEZ- CALCERRADA, 1972	24	Families	Georgia, Puerto Rico, & Dominican Republic	Housing attributes	Differential housing functions and preferences
BETH-HELACHMY, S. & R.L. THAYER, JR., 1978	25	Children	Outdoor play area	Space utili- zation; be- havior patterns	Preferences
BICKMAN, L., A. TEGER, T. CRKRIELE, C. MCLAUGHLIN, M. BERGER, & E. SUN, 1973	26	Students	Dormitory	Density	Helping behavior
BELL, B.D., 1976	27	Elderly	Urban area	Housing type	Satisfaction
BISHOP, D. 1975	28	Pedestrians	England/ street	Environmental setting	Public response
BLAUT, J.M. & D. STEA, 1971	29	Children	Cross- cultural	Geographic learning; cognitive mapping	Map reading photo interpre- tion

AUTHOR YEAR	ABSTR. NO.	MAN	SETTING	RELATIONAL CONCEPTS	OUTCOMES
BOALT, C., 1970	30	Households	Sweden	Household design features	Preferences/ satisfaction
BOOTH, A. & J.N. EDWARDS, 1976	31	Residents	Congested households	Density and crowding	Personality/ family relations
BORN, T.J., 1976	32	Elderly	Campgrounds in Southwest U.S.	Choice of site; differences between groups	Camping preferences and behaviors
BRAND, F.N. & R.T. SMITH, 1974	33	Elderly	Housing	Relocation	Decreased social networks
BRONZAFT, A.L. & D.P. MCCARTHY, 1975	34	Children	Elementary school	Noise; reading skills	Decrease
BROOK, R.M. & P. KNAPP, 1976	35	Children	State residential facility	State-trait anxiety/relocation	Adjustment levels
BROOKES, M.J. & A. KAPLAN, 1973	36	Office workers	Office landscaping	Spatial arrangements	Perception
BROWER, S.N. & P. WILLIAMSON, 1974	37	Residents	Urban neighborhoods	Outdoor recreation	Attitudes/perceptions
BUTLER, E.W., R.J. MCALLISTER, & E.G. KAISER, 1973	38	Adults	Urban areas	Residential mobility	Mental disorder; preferences, intentions
BYROM, C., 1970	39	Residents	Low cost housing, England	Privacy Social interaction, Activity space	Preferences satisfaction
CAHILL, M., 1975	40	Engineering students	Machinery usage	Communication	Symbol recognition and meaning
CAMERON, P., D. ROBERTSON, & J. ZAKS, 1972	41	Families	Urban areas	Sound pollution; noise, and health	Noise judgements
CANTON, D. & R. THONE, 1972	42	Students	Laboratory	Culture	Preferences
CARP, F., 1974	43	Elderly	Residences	Relocation	Person-congruence

AUTHOR YEAR	ABSTR. NO.	MAN	SETTING	RELATIONAL CONCEPTS	OUTCOMES
CARP, F., 1975	44	Elderly	Housing	Hypothetical-actual reloca-tion; defense; cognitive con-sistency	Increased disso-nance associated with moving
CARP, FRANCES, 1977	45	Elderly	Housing	Housing con-ditions	Health
CARR, S.J. & J.M. DABBS, JR., 1974	46	Female students	Laboratory	Distance/ lighting	Behavior
CARTER, D.J. & B. WHITEHEAD, 1976	47	Office workers	Office building/ England	Circulation	Cost-effectiveness communication
CHEYNE, J.A. & M.G. EFRAN, 1972	48	Students/ general public	Shopping mall and university	Territorial intrusion	Use of space
CIMBALO, R.S. & P. MOUSAN, 1975	49	Banking customers	Bank	Crowding	Satisfaction
CIOLEK, M.T., 1977	50	Pedestrians	Urban plaza	Proxemics/ behavior settings	Space usage
CLARK, R.N., J.C. HENDEE & R.L. BURGESS, 1972	51	Human	Theater/ Campground	Reinforced incentives	Behavioral adjustment
CLARK, R.J., 1974	52	Families	Households/ England	Noise	Avoidance reac-tion aural pre-ference environ-mental design
COHEN, S., D.C. GLASS & J.E. SINGER, 1973	53	Children	Apartments	Noise	Reading ability
COREY, J. & C.D. HAMAD, 1976-77	54	Students	Cafeteria	Reinforce-ment signage	Behavioral change
CORLETTE, E.N. & C. HUTCHENSON, 1972	55	Human	Stairs and Ramps	Physiology	Conservation of energy
CORSON, J.H. & C.W. CRANNELL, 1970	56	Students	Laboratory	Tactile thermal response	Warm-cold response vari-ation

AUTHOR YEAR	ABSTR. NO.	MAN	SETTING	RELATIONAL CONCEPTS	OUTCOMES
CRUMP, S.L., D.L. NUNES, & E.K. CROSSMAN, 1977	57	Recreation- ists	Park	Presence of litter	Littering behavior
CUTLER, S.J., 1972	58	Elderly	Rural community	Transporta- tion; resi- dential loca- tion; life satisfaction levels	Lack of personal transportation; low life-statis- faction
DAVIDSON, A.I.G., H.G. SMYLIE, A. MACDONALD, & G. SMITH, 1971	59	Patients	Hospital wards	Treatment setting design	Cross-infection control
DAVIES, A.D.M. & M.G. DAVIES, 1971	60	Teachers students	Schools/ England	Thermal comfort solar design	Environmental preference
DAVIES, D.R., L. LANG & V.J. SHACKELTON, 1973	61	Students	Laboratory	Detection latencies, noise, vigi- lance	Visual vigilance
DIXON, N.F. & E.J. HAMMOND, 1972	62	Students	Laboratory	Trace bright- ness, spatial separation field illumination	Visual persistence
DRINKWATER, B.L., P.B. RAVEN, S.M. HORVATH, J.A. GLINER, R.O. RUHLING, N.W. BOLDUAN, & S. TAGUCHI, 1975	63	Healthy young males	Laboratory	Air pollution, aerobic power	Revised air pollution standards
DUNCAN, J.S. JR. 1973	64	Suburbanites	Residential landscapes	Landscape configuration	Social stratification
EDNEY, J.J., 1972	65	Students	Room	Territori- ality	Interpersonal distance
EDWARDS, J.N. & A. BOOTH, 1977	66	Families	Urban	Crowding	Sexual behavior
EISEMON, T., 1975	67	Residents	Public housing	Simulation techniques	Preferences
EISENBERG, T.A. & R.L. McGINTY, 1977	68	Students	Laboratory	Career Orien- tation	Spatial Visualization
ELSON, M.J., 1976	69	Car Owners	England	Recreational areas, preference	Spatial search behavior
EVERETT, M.D., 1974	70	Recrea- tionists	Recreation areas	Recreation pollution	Facility usage

AUTHOR YEAR	ABSTR. NO.	MAN	SETTING	RELATIONAL CONCEPTS	OUTCOMES
EVERITT, J.C., 1976	71	Husbands and wives	Urban area	Propinquity; cognitive mapping	Site recalcitrance; behavior patterns
FANGER, P.O., 1973	72	Students	Laboratory, Denmark	Thermal factors, age, adaptation, sex, circadian rhythm	Thermal comfort, thermal preference
FELTON, B. & E. KAHANA, 1974	73	Elderly	Institution	Locus of control	Positive adjustment
FIEDLER, F.E. & J. FIEDLER, 1975	74	Neighborhood residents	Residential areas near airports	Noise	Attitudes/ observed behaviors/tolerance
FITCH, J.M., J. TEMPLER, & P. CORCORAN, 1974	75	Human	Stairways	Stairs, use of/design of	Safety on stairs
FLOCK, H.R. & K. NOGUCHI, 1973	76	Students	Laboratory	Brightness contrasts	Brightness perception
FLOCK, H.R., K. NOGUCHI, & P.M. MUTER, 1974	77	College students	Laboratory	Lightness	Lightness perception
FOX, D.J., 1972	78	Mexican	Housing	Sanitation	Mortality
FOX, R.H., R. MacGIBBON, L. DAVIES, & P.M. WOODWARD, 1973	79	Elderly	Residence	Thermal comfort Age	Thermal awareness
FOX, R.H., P.M. WOODWARD, A.N. EXTON-SMITH, M.E. GREEN, D.V. DONNISON, & M.H. WICKS, 1973	80	Elderly	Residence	Thermal comfort Age	Hypthermia susceptibility, thermal awareness
FRASER, T.M. & B.E. MOTTERSHEAD, 1976	81	Males	All terrain vehicle	Buffeting/ human response	Effect on body joints
FREEDMAN, J.L., A.S. LEVY, R.W. BUCHANA, & J. PRICE, 1972	82	Human	Rooms	Crowding, sex sex differences	Aggressiveness

AUTHOR YEAR	ABSTR. NO.	MAN	SETTING	RELATIONAL CONCEPTS	OUTCOMES
FREEDMAN, J.L., S. HESHKA, & A. LEVY, 1975	83	Residents	Urban	Density	No effect of density, pathology (disease)
GARDNER, M.B., 1971	84	Human	Restaurant, auditorium, foyer	Acoustics	Verbal communication
GASPARINI, A., 1973	85	Human	House	Environmental arrangement	Social behavior
GELLER, E.S., J.F. WITMER & M.A. TUSCO, 1977	86	Consumers	Supermarket	Littering	Non-littering behavior
GELPERIN, A., 1974	87	Males	Barracks	Humidity control Respiratory infection	Physical health
GOODMAN, R.F. & B.B CLARY, 1976	88	Human	Los Angeles	Exposure to aircraft noise	Political and social response
GOULD, P., 1977	89	Human	Sweden	Mental mapping	perception
GRANDJEAN, E. & W. HUNTING, 1977	90	Workers	Standing and sitting	Posture	Pain
GRAVURIN, E.I., & D. MURGATROYD, 1975	91	University students	Laboratory	Spatial Aptitude	Permutational ability
GROVES, D.L. & H. KAHALAS, 1975	92	Recreationists	Recreation area	Environmental meaning	Image
GUILFORD, J.S., 1973	93	Females	Controlled kitchen environment	Accident behaviors	Accident predictor variables
HALL, R., A.T. PURCELL, R. THOME, & J. METCALFE, 1976	95	Human	Laboratory	Architectural arrangement	Preferences
HAMILTON, P., G.R.J. HOCKEY, & J.G. QUINN, 1972	96	Housewives College students	Sonic environment	Background noise Long-term memory Short-term memory	Recall performance

AUTHOR YEAR	ABSTR. NO.	MAN	SETTING	RELATIONAL CONCEPTS	OUTCOMES
HANEY, W.G. & E.S. KNOWLES, 1978	97	Human	Neighbor-hoods	Size	Cognitive maps perception
HARRISON, J.A. & P. SARRE, 1975	98	Residents; shopkeepers	Urban area	Personal construct theory; rep-ertory grid	Imagery
HARTMAN, C., J. HOROWITZ & R. HERMAN, 1976	99	Elderly	Laboratory	Environmental manipulation	Preferences
HASKELL, B., 1975	100	Aircrews	Aircraft	Noise	Disease
HAWKINS, F., 1974	102	Pilots aircrew	Airplanes	Seat design/ back pain	Design modification
HAYWARD, S.C. & S.S. FRANKLIN, 1974	103	Students	Laboratory	Boundaries	Space perception
HAZELWOOD, M.G. & W.J. SHULDT, 1977	104	Male college students	Laboratory	Physical phenomenolog-ical distance	Self-disclosure
HEFT, H., 1978	105	Children	Home	Noise, archi-tectural arrangements	Attention levels
HENDEE, J.C., R.P. GALE & W.R. CATTON, JR, 1971	106	Human	Parks	Recreation	Preference
HERMAN, J.B. & A.W. SIEGAL, 1978	107	Children	Scale model town	Cognitive mapping of large-scale environments	Accuracy levels via repeated ex-posure
HERMANN, E. & H. CHANG, 1974	108	Train riders	Trains	Noise/health hazard	Hazard determina-tion
HERZOG, T.R., S. KAPLAN, & R. KAPLAN, 1976	109	Students	Laboratory	Urban places	Preferences
HESS, R.F. & N. DRASDO, 1974	110	Epileptics	Environment	Seizures/ flickering light	Preventions & treatment
HIGH, T. & E. SUNDSTROM, 1977	111	Students	Dormitories	Architectural flexibility	Behavior
HILL, M.R. & T.T. ROEMER, 1977	112	Pedestrians	Intersections	Jaywalking behavior	Social context
HOLAHAN, C. & S. SAEGERT, 1973	113	Psychiatric patients	Psychiatric ward	Environmental arrangement	Social behavior
HOLAHAN, C., 1976	114	Human	Urban neigh-borhood	Environmental arrangement	Social behavior
HOLAHAN, C., 1977	115	Students	Laboratory	Urban ex-posure	Helping behavior
HUDSON, R., 1974	116	Consumers	Downtown shopping area	Cognition	Spatial behavior
HUGHES, J. & M. GOLDMAN, 1978	117	Experimental subjects	Elevators	Eye contact, facial ex-pression, sex differences in	Personal space violation

AUTHOR YEAR	ABSTR. NO.	MAN	SETTING	RELATIONAL CONCEPTS	OUTCOMES
HUMPHREY, C., D.A. BRADSHAW, & J.A. KROUT, 1978	118	Upper-middle class household residents	Interstate highway residential area	Effects of air & noise pollution	Reaction to environmental adversities
HUMPHREYS, M.A.,	119	General population	Outdoor microclimate	Thermal factors	Thermal comfort outdoor microclimate design
HUMPHREYS, M.A., 1977	120	School children	Primary school, England	Thermal factors	Methodology revision, thermal comfort
HUNT, J.C.R., E.C. POULTON, & J.C. MUMFORD, 1976	121	Human	Wind tunnel	Wind	Comfort
IZUMI, K., 1976	122	Mentally ill	Clinic space	Environmental manipulation	Psychological response
JACOBSEN, C., S. HAVEH, & T. AVI-ITZHAK, 1975	123	Israelis	Kibbutzim	Social control	Environmental cleanliness
JONES, J.W. & A. BOGAT, 1978	124	Students	Room	Air pollution, aggression	Increased aggressiveness in polluted conditions
JONES, F.N., D. NOWELL, & J. TAUSCHER, 1978	125	Residents	Airport landing area	Noise, pollution	Mental and physical health
KEIGHLEY, E.C., 1973	127	Office workers	Laboratory	Aesthetic quality fenestration	Visual preference
KEIGHLEY, E.C., 1973	128	Office workers, English	Controlled setting	Aestetic quality fenestration	Visual preference
KAHN, M.A.Q., W.F. COELLO, & Z.A. SALEM, 1973	129	Man	Expressway	Proximity/ lead content of soil	Lead contamination
KIMBER, C.T., 1973	130	Puerto Ricans	Entry gardens	Terrioriality	Social stratification
KLASS, D.B., G.A. GROWE, & M. STRIZICH, 1977	131	Psychiatric patients	Psychiatric hospital	Social structure, hostile impulse control, time-space structuring	Adjustment, recidivism rates
KNAVE, B., H.E. PERSSON, M. GOLDBERG, & P. WESTERHOLM, 1976	132	Workers	Factory	Air pollution, health	Disease

AUTHOR YEAR	ABSTR. NO.	MAN	SETTING	RELATIONAL CONCEPTS	OUTCOMES
KOEPPEL, J.C. & K.W. JACOBS, 1974	133	College students	Laboratory	Past mobility, sensation seeking, extraversion	Mobility
KONEYA, M., 1976	134	Students	Classroom	Seating arrangement	Social interaction (verbal behavior)
KORTE, C. & N. KERR, 1975	135	Human	Urban	Environment	Altruistic response
KOSHAL, R.K. & M. KOSHAL, 1974	136	Human	United States	Air pollution	Disease (health)
KRAUSS, R.M., J.L. FREEDMAN, & M. WHITCUP, 1978	137	Students, pedestrians	Urban setting, laboratory	Situational variables	Littering behavior
KULDAU, J.M. & S.J. DIRKS, 1977	138	Psychiatric patients	Psychiatric hospital, community housing	Patient self-management, group therapy, planning	Adjustment
LAIRD, J.T., 1973	139	Human	Urban areas	Population density	Mental health
LANGER, E.J. & S. SAEGERT, 1977	140	Shoppers	Supermarkets	Crowding	Performance
LAVE, L. & E. SESKIN, 1972	141	Man	Urban environment	Pollution/health	Pollution abatement
LAWTON, F.G. & T.V. BUSSE, 1972	142	Students	Classroom	Residential mobility	Creativity
LAWTON, M.P. & J. COHEN, 1974	143	Elderly	Housing	Relocation	Well-being
LAWTON, P.M., B. LIEBOWITZ, & H. CANTON, 1970	144	Senile elderly	Geriatric center	Remodeling/behavior	Behavioral changes
LAWTON, M.P., M.H. KLEBAN, & M. SINGER, 1971	145	Elderly Jewish	Inner city neighborhoods	Housing, social economic conditions	Deprivations
LAWTON, M.P., 1976	146	Elderly	Housing	Housing arrangements	Satisfaction
LEADERER, B.P., R.T. ZAGRANISKI, & J.A.J. STOLVIJK, 1976	147	Human	United States	Nitrogen dioxide pollution	Health
LECUYER, R., 1976	148	Human	Tables	Spatial organization	Social interaction

AUTHOR YEAR	ABSTR. NO.	MAN	SETTING	RELATIONAL CONCEPTS	OUTCOMES
LENNQUIST, S., P.O. GRANBERG, & B. WEDIN, 1974	149	College students	Thermal environment	Thermal exposure	Physiological changes
LEFF, H.L., L.R. GORDON, & J.G. FERGUSON, 1974	150	Students	Street views	Cognitive sets	Perception
LEVINE, R., 1976	151	Employees neighboring residents	Scrap smelter	Occupational lead poisoning/environmental contamination	Health hazard
LINN, M.W., EM. CAFFEY JR., C.J. KLETT, & G. HOGARTY, 1977	152	Psychiatric patients	Hospital, foster care homes	Patient self-management	Adjustment
LIPMAN, A. & R. SLATER, 1977	153	Elderly	Homes for the elderly	Seating arrangements	Spatial & social organization
LUDWIG, A.M., 1971	154	Human	Laboratory	Sensory deprivation, sensory overload	Environmental preferences
LUNDEN, G., 1972	155	Office workers	Office buildings, Sweden	Work environments	Job satisfaction
MACDONALD, W.A. & E.R. HOFFMAN, 1973	156	Students	Actual & simulated roadways	Threshold recognition	Roadway lettering effectiveness formulas
MAITREYA, V.K., 1977	157	Male	Buildings, India	Gloom, lighting perference, performance	Artificial lighting standards
MARGULIS, S.T., 1977	158	Diverse user groups	Recent studies	Privacy	Summary of current privacy concepts
MARKSON, E. & J.H. CUMMING, 1974	159	Elderly	Hospitals	Involuntary relocation	Mortality
MARKUS, E., M. BLENKNER, M. BLOOM, & T. DOWNS, 1972	160	Two elderly groups	Institution	Relocation, physical status, mortality, coping, mental status	Undifferentiated mortality rates
MARSHALL, N.J., 1972	161	Students, parents	Suburbs	Environmental surroundings	Perceived privacy

AUTHOR YEAR	ABSTR. NO.	MAN	SETTING	RELATIONAL CONCEPTS	OUTCOMES
MATHER, C.E., 1973	162	Communities contiguous to airports	Sonic environment	Noise	Noise susceptibility
MATHEWS, K.E. & L.K. CANON, 1975	163	Physically disabled	Outdoor environment	Noise levels, helping behavior	Interpersonal relationships
MAURER, R. & J.C. BAXTER, 1972	164	Children	Ethnically mixed neighborhood	Cognitive maps, perception	Map utility
MAYRON, L.W., E.L. MAYRON, J. OTT, & R. NATIONS, 1974	165	Children	Schools	Flourescent lighting	Behavior, Dental caries
MAWBY, R.I., 1977	166	Victims of crimes/ police files	Residential areas, England	Defensible space	Physical layout, importance of high-rise developments, problems of gardens, use of
MACALLISTER, R.J., E.W. BUTLER, & E.J. KAISER, 1973	167	Women	Residential households	Mobility and adaptation	Coping behavior/ adjustment
MCCARTHY, D. & S. SAEGERT, 1978	168	Apartment dwellers	Apartments	Density	Social overload, withdrawal
MCKAIN, J.L., 1973	169	Army families	Residential portions of army	Relocation/ alienation/ maladjustment	Alienated wife-mother/community isolation
MEADE, S., 1977	170	Malaysians	Agricultural settlers	Migration	Morbidity
MEHRABIAN, ALBERT & SHIRLEY G. DIAMOND, 1971	171	University students	Laboratory	Seating proximities; affiliative tendency	Preferences
MICHELSON, W., D. BELGUE, & J. STEWART, 1973	172	Families	Urban area/ housing	Differential residential selection	Intentions; preferences; expectations; actual behaviors
MITCHELL, J.K. 1976	173	Suburban immigrants	Suburbs	Environmental clues	Perceptions
MITCHELL, R.E. 1971	174	Adults; children	Urban area	Overcrowding	Relationships; effects
MORGAN, J.C., J.S. LOCKARD, C.E. FAHRENBRUCH, & J.L. SMITH, 1975	175	Automobile drivers	Roadside	Communication	Assistance offered hitchhikers

AUTHOR YEAR	ABSTR. NO.	MAN	SETTING	RELATIONAL CONCEPTS	OUTCOMES
MORRIS, E.W., S.R. CRULL, & M. WINTER, 1976	176	Families	Metropolitan area/housing	Satisfaction/propensity to move	Normative housing deficit model
MORRIS, E.W., 1977	177	Females	Rural villages	Crowding	Fertility behavior/residential mobility
MYSKA, M.J. & R.A. RASEWARK, 1978	178	Elderly	Rural institutional/non-institutional	Death attitudes	No differentiation
NE'EMAN, E., W. LIGHT & R.G. HOPKINSON, 1976	179	Human	Buildings	Sunlight exposure	Visual preference; design revision
NE'EMAN, E., J. CRADDOCK, & R.G. HOPKINSON, 1976	180	Office workers; general population	Residence, school, office, hospital	Sunlight penetration	Visual sunlight
NE'EMAN, E., 1977	181	General population	Visual, school, hospitals, universities	Sun penetration, thermal	Visual preference, thermal comfort
NOGAMI, G.Y., 1976	182	Human	Rooms	Crowding	Perception
NORMAN, K.L., 1977	183	Students	Room	Bus system effectiveness	Attribute judgement model
O'LEARY, K.D. & A. ROSENBAUM, 1978	184	Children	Classroom	Flourescent lighting	Hyperactive behavior
OLOKO, O., 1973	185	Factory workers	Nigeria	Environmental setting	Worker commitment
OLSEN, H. & R. CARTER, 1974	186	Children	Urban/rural	Location; learned ability	Grade level-setting influences learning
OLSHAVSKY, R.W., D.B. MACKAY, & G. SENTALL, 1975	187	Consumers	Urban area	Perceptual/actual distance	Compared perceptions
OSTBERG, O. & A.G. MCNICHOLL 1973	188	College students	Controlled thermal	Thermal factors; circadian rhythm	Thermal preference variation
PAGE, R.A., 1977	189	Human	Street	Noise	Helping behavior
PALM, R., 1976	190	Real estate agents	Neighborhoods	Perceptions	Preferences

AUTHOR YEAR	ABSTR. NO.	MAN	SETTING	RELATIONAL CONCEPTS	OUTCOMES
PAMPEL, F.C. & H. M. CHALDIN 1978	191	Elderly, urban resi- dents	Urban areas	Segregation/ integration	Ecological models
PAULUS, P., V. COX, G. MCCAIN, & J. CHANDLER, 1975	192	Prison inmates	Single cells; dormitories	Crowding	Negative consequences
PEDERSON, D.M. 1977	193	University students	Laboratory	Size of home town	Perception of environment
PERETTI, P.O. 1977	194	Elementary students	Laboratory	Illumination	Verbal response
PETERSON, E.T. 1973	195	Residents	Freeway environs	Proximate/ attitudes	Self-image
PETERSON, G.L., BISHOP, R.L., MICHAELS, R.M. & RATH, G.J., 1973	196	Children	Playground	Equipment preferences	Scaling tech- nique/design
PLANEK, T.W. & FOWLER, R.C., 1971	197	Elderly drivers	Urban-rural; two states	Frequency; exposure; content	Differences; judgements
POPENDORF, W. & SPEAR, R., 1974	198	Harvesters	Fields	Pesticides/ exposure	Dust exposure level
POULTON, E.C., HUNT, J.C.R., MUMFORD, J.C. & POULTON, J., 1975	199	Women	England	Wind effects	Task decrement
PUTZ, V. & K.U. SMITH, 1971	200	Machine operators	Simulated work environment	Retinal feedback	Movement capability
RAE, A. & R.M. SMITH, 1976	201	Hospital patients	Surgical ward	Ventilation rate	Odor perception
REA, W.J., 1976	202	Thrombo- phlebitis patients	Hospital	Treatment settings	Health management
REA, W.J., 1978	203	Cardiac patients	Hospital	Treatment setting	Health management
REES, C.A., 1975	204	Elderly	Urban	Size and growth rate	Age segregation
REICHNER, R.F. 1979	205	College freshmen	Dormitories	Spatial arrangement	Social isolation
RICHARDS, O., 1977	206	16 to 90 year olds	Room	Luminance & visual acuity	Optimal levels

AUTHOR YEAR	ABSTR. NO.	MAN	SETTING	RELATIONAL CONCEPTS	OUTCOMES
RILEY, R.L., 1972	207	Hospital patients	Hospital	Airborne infection; indoor air pollution	Reduced infection, air disinfection device
RODIN, J. & E.J. LANGER, 1977	208	Elderly	Nursing homes	Control; responsibility	Lower mortality rates
ROHNER, R.P., 1974	209	Male students	Dormitory	Crowding	Stress
RONALD, P.Y., M.E. SINGER & F. FIREBAUGH, 1971	210	Homemakers	Residences	Task dimensions	Preferences
ROOT, J.D., 1975	211	College students	Laboratory	Intransitivity	Environmental preferences
ROTHBLATT, D.N. 1971	212	Adults/ general population	Public housing	Human needs/ housing	Adjustment
SAEGART, S., MACKINTOSH, E. & WEST, S., 1975	213	Shoppers/ commuters	Department store, terminal	Crowding, stress, cognition	Low performance, crowded conditions
SCHAEFER, C.E. & J. HIGGINS, 1976	214	Children	Institution	Interpersonal distance; sociometric status	Preferences
SCHIFFENBAUER, A.I., J.E. BROWN, P.L. PERRY, L.K. SCHULACK, & A.M. ZANZOLA, 1977	215	Female students	Dormitory	Architectural arrangement	Perceived crowdedness
SCHMIDT, D.E., R.D. GOLDMAN, & N.R. FEIMER, 1976	216	Three ethnic groups	Urban area	Crowding/ density	Perceptions/ attitudes
SCOTT, J., 1970	217	Residents	Urban	Design, social theory, life satisfaction	Housing assessment criteria
SHAPIRO, M.T., W. MELNICK, & V. VERMEULEN, 1972	218	Middle-aged males, hearing impaired males	Audially controlled room	Noise, hearing loss	Performance
SHERMAN, S.R., 1974	220	Elderly	Retirement housing	Leisure activities	Adjustment
SHERMAN, S.R., 1975	221	Elderly	Six retirement housing sites	Mutual assistance support networks; relocation	Patterns of interdependence

AUTHOR YEAR	ABSTR. NO.	MAN	SETTING	RELATIONAL CONCEPTS	OUTCOMES
SILVERSTEIN, C.H. & J. STANG, 1976	222	Human	Tables	Seating position	Social interaction
SIMS, B. & S. KAHN, 1977	223	Residents	Urban area	Exposure; uniqueness; significance	Predictive model; landmark imageability
SLIWOWSKI, L., 1972	224	Inhabitants	Apartments	Microclimate, environment, thermal	Comfort criteria revision
SMITH, B.W. & J. HILTNER, 1975	225	Elderly	Residences	Intraurban location	Distribution
SMITH, J.M., W.T. OSWALD, & G.Y. FARUKI, 1976	226	Psychiatric inpatients	Hospital	Involuntary relocation	Adjustments, satisfaction
SMYLIE, H.G., A.I.G. DAVIDSON, A. MACDONALD & G. SMITH, 1971	227	Injured patients	Hospital wards	Treatment setting design	Cross-infection control
SNYDER, R. & WILLIAMS, A., 1973	228	Students	University	Transportation, housing	Public transportation, needs for
SPINETTA, J.J., D. RIGLER, & M. KARON, 1974	229	Children	Hospital	Personal space	Preferred psycho-social distances
STOKOLS, D, R.W. NOVACO, J. STOKOLS, & J. CAMPBELL, 1978	230	Commuters	Simulated highway trips	Stress judgements, trip type	Impedance levels, physiological consequences
STOKOLS, D. W. OHLIG, & S.M. RESNICK, 1978	231	College students	Residences, classrooms	Crowding	Health
STORANDT, M. I. WITTELS, & J. BOTWINICK, 1975	232	Elderly	Residential	Dimension of well-being, relocation	Successful adjustadjustment, pre-
STORANDT, M. & I. WITTELS, 1975	233	Elderly	Institutional/non-institutional residential setting	Relocation	Adjustment
STUTZ, F.P., 1976	234	Elderly males	Hotel	Life support networks	Disruption
TAINSH, M.A., 1975	235	Travellers	Car, train	Travel frequency	Health status

AUTHOR YEAR	ABSTR. NO.	MAN	SETTING	RELATIONAL CONCEPTS	OUTCOMES
TAYLOR, L.H. & E.W. SOCOV, 1974	236	Human	Laboratory	Lighting levels	Mobility prefer- ences
TAYLOR, L.H., E.W. SOCOV, & D.H. SHAFFER, 1974	237	Human	Laboratory	Lighting	Preferences
TEAFF, J.D., M.P. LAWTON, L. NAHEMOW & D. CARLSON, 1978	238	Elderly	Public housing	Age inte- gration and well-being	Positive impact, age-segregation
TENNIS, G.H. & J.M. DOBBS, 1975	239	Students	Urban schools	Age/sex differences, interper- sonal distance	Age-related interpersonal distance
TOGNOLI, J., 1973	240	Students	Laboratory	Environ- mental setting	Behavior
TUCKER, J., 1972	241	Students	Rooms	Density interaction	Greater density, decreased inter- action
Tyler, D.A., 1973	242	Truck drivers	Trucks	Design/ noise	Noise reduction
TYNDER, L., 1973	243	Nursing home patients	Nursing homes	Design/ social interaction	Optimal interac- tion
ULRICH, R.S., 1977	244	Students, suburanities	Sweden, U.S.	Cross- cultural, preference model	Similarities, dimensions
VALINS, S. & A. BAUM, 1973	245	Students	Laboratory	Group size	Social interac- tion
VANSELOW, G.W., 1974	246	Random subjects	University district	Horizontal and vertical components	Spatial changes
VERDERBER, S. & G.T. MOORE, 1977	247	Urban dwel- lers, stud- ents, subur- banites	Urban area	Comparative building imagery, cognition	Dimensions, pre- ferences
WALLER, R.A., 1975	248	Human	Living environments	Measurement and annoy- ance	Environmental quality
WARNER, P., J. JACKSON, L. GOOD, & A. GREENBURG, 1975	249	Man	Residential	Effluents/ pollution	Effluents effect

AUTHOR YEAR	ABSTR. NO.	MAN	SETTING	RELATIONAL CONCEPTS	OUTCOMES
WEINSTEIN, N.D., 1978	250	University students	Dormitory	Noise/ longitudinal analysis	Sensitivity judgements
WHEELER, J.O. & F.P. STUTZ, 1971	251	Residents	Urban area	Distance decay, social trips, socioeco- nomic status	Social trip patterns
WHITE, S.E., 1974	252	Urban dwellers	Cities	Residential preferences	Urban in-migra- tion
WHITEHEAD, B., 1970	253	School staff, students	Secondary school	Design determinants, routine journeys	Design recommen- dations
WINETT, RICHARD A., 1977-79	254	Students	Classrooms	Reinforce- ment signage	Unambiguous cues
WINKLER, L.H., 1978	255	Space colonizers	Space colonies	Adaptation/ space environment	Physiological constraints
WITTELS, I. & J. BOTWINICK, 1974	256	Elderly	Housing	Relocation, mortality rates	Positive influ- ence
WOHLWILL, J.F., J.L. NASAR, D.M. DEDOY & H.H. FORUZANI, 1976	257	University students	Room	Noise exposure	Activity patterns/toler- ance
WOLFE, M., 1975	258	Children	Psychiatric hospital	Room size, group size, density	Density factors
WOLGERS, B., 1973	259	Employees	Office buildings	Attitudes	Work setting preferences
WOLK, S. & S. TELLEEN, 1976	260	Elderly	Residential	Constraint, satisfaction	Low constraint preferences
WOOLS, R. & D. CANTER, 1970	261	Housewives	Scotland	Meaning, behavior	Perception
WORTHINGTON, M.E., 1974	262	Passersby	Airport lobby	Personal space	Stigma effect
YEE, W. & M.D. VANARSDOL, 1977	263	Families	Residences	Mobility, age, life cycle	Predictive mobility model

SETTING INDEX	ABSTRACT NUMBER
Elevators	117
England	69, 199
Entry garden	130
Environment	110
Environmental arrangement	4
Expressway	129
Factory	11, 132
Fields	198
Football stadium	16
Foyer	84
Freeway environs	195
Georgia	24
Geriatric center	144
Highway trips, simulated	230
Home	105
Hong Kong	101, 174
Homes for the elderly	153
Hospital wards	59, 227
Hospital	94, 152, 159, 180, 202, 203, 207, 226, 229
Hospital laboratory	10
Hospital, psychiatric	21, 131, 138, 258
Hotel	234
House	85
Households	52
Housing	23, 33, 39, 44, 45, 78, 85, 143, 146, 172, 176, 212, 238, 256
Indoors	7
Inner city neighborhoods	145

RELATIONAL CONCEPTS	ABSTRACT NUMBER
Acoustics	84
Activity space	39
Adaptation/space environment	255
Aesthetic quality	127, 128
Age	72, 79, 80, 264
Age integration	238
Age/sex differences	239
Aggression	124
Aircraft noise, exposure to	88
Air pollution	124, 136
Air pollution, aerobic power	63
Air pollution, health	132
Alienation	169
Annoyance, measurement of	248
Architectural arrangement	23, 95, 215
Architectural elements	12
Architectural flexibility	111
Attitudes	259
Back pain	102
Behavior	262
Behavior patterns	25
Behavior settings	50
Boundaries	103
Brightness contrasts	76
Building material, pollution	7
Building materials	15
Bus system effectiveness	183

RELATIONAL CONCEPTS	ABSTRACT NUMBER
Career orientation	68
Choice of site	32
Circadium rhythm	72, 188
Circulation	47
Cognition	116, 213, 247
Cognitive mapping of large-scale environments	107
Cognitive maps, perception	29, 71, 164
Cognitive sets	150
Collective decisions	18
Communication	40, 47, 175
Comparative building imagery	247
Constraint, satisfaction	260
Control	208
Coping	160
Cross-cultural, preference model	244
Crowding	2, 19, 31, 49, 66, 82, 101, 140, 177, 182, 192, 209, 213, 216, 231
Culture	42
Death attitudes	178
Defensible space	166
Densely populated environment/ social behavior	14
Density	5, 26, 31, 83, 168, 216, 241
Design	217, 242, 243, 253
Detection latencies, noise, vigilance	61
Differential residential selection	172
Distance	46, 187
Distance/lighting	46

RELATIONAL CONCEPTS	ABSTRACT NUMBER
Effluents	249
Ego-defense	44
Environment	135
Envrionment, densely populated	14
Environmental	22, 85, 113, 114
Environmental clues	173
Environmental control	126
Environmental manipulation	99, 122
Environmental meaning	92
Environmental setting	28, 185, 240
Environmental surroundings	161
Equipment preferences	196
Exposure; uniqueness; significance	197, 223
Eye contact, sex differences in	117
Field illumination	62
Flourescent lighting	165, 184
Geographic learning	29
Geographic variables	17
Group cohesiveness	13
Group differences	32
Group size	245
Group therapy	138
Growth rate	204
Health	41, 141
Health hazard	108
Hearing loss	218
Helping behavior	163
High altitude, prolonged stay	219

RELATIONAL CONCEPTS	ABSTRACT NUMBER
Horizontal components	246
Hostile impulse control	131
Household design features	30
Housing	212, 228
Housing arrangements	146
Housing attributes	24
Housing conditions	45
Housing density	101
Housing type	27
Humidity control-respiratory infection	87
Illumination	10, 194
Infection, airborne	207
Indoor air pollution	207
Interpersonal distance; socio-metric status	214
Intrasitivity	211
Intraurban location	225
Involuntary relocation	94, 159, 226
Jaywalking behavior	112
Kitchen accident behaviors	93
Landscape configuration	64
Lead content of soil	129
Leisure activities	220
Life cycle	263
Life satisfaction levels	58, 217
Life support networks	234
Light, flickering	110
Lighting	46, 237

RELATIONAL CONCEPTS	ABSTRACT NUMBER
Lighting levels	236
Lighting preference, performance	157
Lightness	77
Litter, presence of	57
Littering control	16, 86
Location; learned ability	186
Locus of control	73
Luminance & visual acuity	206
Maladaptation; social indicators, geographical variables	17
Meaning, behavior	261
Memory, long term/short term	96
Mental mapping	89
Microclimate, environment, thermal	224
Migration	170
Mobility and adaptation	167
Mobility, age, life cycle	263
Mobility, past	133
Mutual assistance support networks; works; relocation	221
Noise	9, 52, 53, 61, 74, 100, 162, 189, 242
Noise, background	96
Noise exposure	257
Noise levels, helping behavior	163
Noise, architectural arrangements	105
Noise, hearing loss	218
Noise/health hazard	108
Noise/illumination	10

RELATIONAL CONCEPTS	ABSTRACT NUMBER
Noise/longitudinal analysis	250
Noise; reading skills	34
Occupation	11
Occupational lead poisoning/environmental contamination	151
Overcrowding	174
Patient self-management	152
Patient self-management, group	138
Perceptions	190
Perceptual/actual distance	187
Personal construct theory; repertory grid	98
Personal space	20, 229, 262
Pesticides/exposure	198
Physical phenomenological distance	104
Physiology	55
Pollution, air, noise	118, 125
Pollution/building material	7
Pollution, health	141, 249
Pollution, nitrogen dioxide	147
Population density	139
Posture	90
Privacy	8, 39, 158
Propinquity; cognitive mapping	71
Proxemics	1, 50
Proximate/attitudes	195
Recreation	106
Recreation, outdoor	37
Recreation pollution	70

RELATIONAL CONCEPTS	ABSTRACT NUMBER
Recreational areas, preference	69
Reinforced incentives	51
Reinforcement signage	54, 254
Relocation	33, 35, 43, 143, 232, 233
Relocation, hypothetical-actual	44
Relocation, mortality rates	256
Relocation, physical status, mortality, coping, mental status	160
Relocation/alientation/maladjust-ment	169
Remodeling/behavior	144
Residential mobility	38, 142
Residential preferences	252
Respiratory infection	87
Retinal feedback	200
Room size, group size, density	258
Routine journeys	253
Sanitation	78
Satisfaction/propensity to move	176
Seat design/back pain	102
Seating arrangement	134, 153
Seating position	222
Seating proximities	171
Segregation/integration	191
Seizures/flickering light	110
Sensory deprivation, sensory over-load	154
Sex	72
Simulation techniques	67

RELATIONAL CONCEPTS	ABSTRACT NUMBER
Situational variables	137
Size and growth rate	97, 204
Size of home town	193
Social control	123
Social interaction	243
Social, socioeconomic status	251
Social structure, hostile impulse control, time-space structuring	131
Sound pollution; noise, and health	41, 125, 141
Space utilization; behavior patterns	25
Spatial Aptitude	91
Spatial arrangements	36, 205
Spatial organization	148
Stairs, use of/design of	75
Stress judgements, trip type	230
Sun penetration	180, 181
Sunlight exposure	179
Tactile thermal response	56
Task dimensions	210
Territorial intrusion	48
Territoriality	4, 65, 130
Thermal comfort	3, 6
Thermal factors	119, 120
Thermal factors building materials	15
Threshold recognition	156
Trace brightness, spatial separation, field illumination	62
Transportation, housing	58, 228

RELATIONAL CONCEPTS	ABSTRACT NUMBER
Travel frequency	235
Treatment setting design	59, 227
Treatment settings	21, 202, 203
Urban exposure	115
Urban places	109
Ventilation rate	201
Wind	121
Wind effects	199
Work environments	155

OUTCOMES	ABSTRACT NUMBER
Accident predictor variables	93
Accuracy levels via repeated exposure	107
Activity patterns/tolerance	257
Adaptation	94
Adjustment	43, 73, 138, 152, 167, 212, 220, 226, 232, 233
Adjustment levels	35
Adjustment, recidivism rates	131
Age segregation	204, 238
Age-related interpersonal distance	239
Aggressiveness	82, 124
Air pollution standards, revised	63
Altruistic response	135
Artificial lighting standards	157
Attention levels	105
Attitudes/observed behaviors/ tolerance	74
Attitudes/perceptions	37
Attribute judgement model	183
Avoidance reaction aural preference environmental design	52
Behavior	46, 111, 165, 240
Behavior modification	21
Behavioral adjustment	51
Behavioral change	54, 144
Body joints, effects on	81
Brightness perception	76
Camping preferences and behaviors	32

OUTCOMES	ABSTRACT NUMBER
Children's behavior	23
Comfort	121
Comfort criteria revision	224
Communication/performance	4
Community isolation	169
Comparative response levels	10
Compared perceptions	187
Conservation of energy	55
Contamination	7
Coping behavior/adjustment	167
Cost-effectiveness	47
Creativity	142
Cross-infection control	59, 227
Crowded conditions	213
Crowdedness, perceived	215
Decrease	34
Density effect of pathology	83
Density factors	258
Deprivations	145
Design modification	102
Design recommendations	253
Differences; judgements	197
Differential housing functions and preferences	24
Differentiation, no	178
Dimensions preferences	247
Disease	100, 132, 136
Disruption	234

OUTCOMES	ABSTRACT NUMBER
Dissonance associated with moving	44
Distribution	225
Dust exposure level	198
Ecological models	191
Effluents effect	249
Environmental adversities, reaction to	118
Environmental analysis	8
Environmental cleanliness	123
Environmental preference	60, 154, 211
Environmental quality	248
Facility usage	70
Fertility behavior/residential mobility	177
Grade level-setting influences learning	186
Hazard determination	108
Health	45, 147, 231
Health hazard	151
Health management	202, 203
Health status	235
Helping behavior	26, 115, 189
Hitchhikers, assistance offered	175
Housing assessment criteria	217
Humidity sensations	6
Hyperactive behavior	184
Hypothermia susceptibility, thermal awareness	80
Image	92

OUTCOMES	ABSTRACT NUMBER
Imagery	98
Impedance levels, physiological consequences	230
Infections, reduced	207
Intentions; preferences; expections; actual behaviors	172
Interaction	241, 243
Interpersonal distance	65
Interpersonal relationships	163
Invasion	20
Job satisfaction	155
Landmark imageability	223
Lead contamination	129
Learning performance	3
Lightness perception	77
Littering behavior	57, 137
Low constraint preferences	260
Map reading, photo interpretation	29
Map utility	164
Measurement scale	126
Mental and physical health	125
Mental disorder; preferences, intentions	38
Mental health	139
Mobility	133
Mobility preferences	236
Morbidity	170
Mortality	78, 159
Mortality rates	94, 160, 208

OUTCOMES	ABSTRACT NUMBER
Movement capability	200
Negative consequences	192
Neighboring, low	14
Noise judgements	41
Noise reduction	242
Noise susceptibility	162
Non-littering behavior	16, 86
Normative housing deficit	176
Odor perception	201
Optimal levels	206
Pain	90
Patterns of interdependence	221
Perception	36, 89, 97, 150, 173, 182, 261
Perception of environment	193
Perceptions/attitudes	216
Performance	13, 140, 218
Permutational ability	91
Person-congruence	43
Personal space violation	117
Personality/family relations	31
Physical health	87
Physical layout, importance of highrise developments, problems of gardens, use of	166
Physiological changes	149
Physiological constraints	255
Political and social response	88
Pollution abatement	141

OUTCOMES	ABSTRACT NUMBER
Positive influence	256
Predictive mobility model	263
Preferences	12, 25, 30, 39, 42, 67, 95, 99, 106, 109, 171, 190, 210, 214, 237
Preventions & treatment	110
Privacy concepts, summary of	158
Privacy, perceived	161
Psychological response	122
Psychomotor, efficiency, impaired	219
Psycho-social distances, preferred	229
Public response	28
Public transportation, needs for	228
Reading ability	53
Recall performance	96
Relationships; effects	174
Residential mobility	177
Roadway lettering effectiveness formulas	156
Safety on stairs	75
Satisfaction	27, 49, 146
Scaling technique/design	196
Self-disclosure	104
Self-image	195
Sensitivity judgements	250
Sexual behavior	66
Similarities, dimensions	244
Site recalcitrance; behavior patterns	71
Social alienation	5

OUTCOMES	ABSTRACT NUMBER
Social behavior	1, 22, 85, 101, 113, 114
Social context	112
Social interaction	134, 148, 222, 245
Social isolation	205
Social networks, decreased	33
Social stratification	64, 130
Social trip patterns	251
Space perception	103
Space usage	48, 50
Spatial & social organization	153
Spatial arrangements	18
Spatial behavior	116
Spatial changes	246
Spatial search behavior	69
Spatial visualization	68
Stigma effect	262
Strain	11
Stress	2, 209
Symbol recognition and meaning	40
Task decrement	199
Task performance	9
Thermal awareness	79
Thermal comfort	15, 72, 119, 120, 181
Thermal preference	188
Transportation, lack of personal	58
Unambiguous cues	254
Urban in-migration	252

DATE DUE